Theory of Elasticity

Yongtao Lyu • Yonggang Zheng

Theory of Elasticity

Basic Concepts and Formulations

Springer SCIENCE PRESS

Yongtao Lyu
School of Mechanics and Aerospace Engineering
Dalian University of Technology
Dalian, Liaoning, China

Yonggang Zheng
School of Mechanics and Aerospace Engineering
Dalian University of Technology
Dalian, Liaoning, China

ISBN 978-981-95-6042-4 ISBN 978-981-95-6043-1 (eBook)
https://doi.org/10.1007/978-981-95-6043-1

Jointly published with Science Press
The print edition is not for sale in China (Mainland). Customers from China (Mainland) please order the print book from: Science Press.?

This Springer imprint is published by the registered company Springer Nature Singapore Pte Ltd.
The registered company address is: 152 Beach Road, #21-01/04 Gateway East, Singapore 189721, Singapore

Preface

The textbook *Theory of Elasticity: Basic Concepts and Formulations* is to be used by the upper-class undergraduate students and the graduates in the fields of engineering mechanics, mechanical engineering, civil engineering, etc. The primary aim of the textbook is to provide the basic concepts and formulations related to the theory of elasticity, including the strain and stress tensors, the constitutive law, the principal strains and principal stresses, the plane stress and plane strain problems, etc. The theory of elasticity is an important basic subject for engineering students and serves as the basis for many advanced engineering courses, such as finite element method, the elastic-plastic mechanics, the continuum mechanics, etc. Therefore, understanding the basic concepts, the relevant formulations and mathematical preliminaries are important and crucial for students to understand the subsequent mechanics courses. It should be noted that this textbook only covers the basic concepts and formulations in elasticity. For some advanced topics, such as the general theory of cylindrical shells, the bending problem of thin plates, variational method, the large deflection problems of thin plates, etc., the readers can refer to other textbooks.

Dalian, China — Yongtao Lyu
Dalian, China — Yonggang Zheng
30 August, 2025

Contents

About the Authors

Yongtao Lyu is an Associate Professor in the School of Mechanics and Aerospace Engineering, Dalian University of Technology, China. Dr. Lyu obtained his Bachelor and Master of Engineering degrees (2001–2007) from Huazhong University of Science and Technology, China. Then he received the EPSRC scholarship and completed Ph.D. study (2007–2010) at Cardiff University, UK. Afterward, he worked as a Postdoctoral Researcher (2010–2013) at the Institute of Biomechanics, Hamburg University of Technology, Germany, and then as a Senior Research Associate (2013–2015) at the Institute for in silico Medicine, the University of Sheffield, UK. At the end of 2015, he started the faculty position at Dalian University of Technology, China, and worked until now. Dr. Lyu's research interests lie primarily in the mechanical behavior of biomaterials, the computational modeling of human body systems and the bionic design of bone scaffolds. He has published over 50 SCI-indexed academic papers. Most of them are in the top-level journals in the field of biomechanics.

Yonggang Zheng is a Full Professor and Vice Dean in the School of Mechanics and Aerospace Engineering, Dalian University of Technology, China. Prof. Zheng obtained his bachelor and Ph.D. degrees from Dalian University of Technology. He then worked as Lecturer, Associate Professor, and Professor in Dalian University of Technology. His main research areas are computational soft matter mechanics, computational nano- and biomechanics, computational impact dynamics, etc. He has published over 130 SCI-indexed papers in top journals in the fields of computational mechanics and solid mechanics (e.g., CMAME, IJNME, ACS Nano, APL, etc.), and received many awards, including the national "Ten Thousand Talents Plan" Young Talent awards, the Ministry of Education's New Century Excellent Talents Program, etc.

Chapter 1
Introduction and Mathematical Preliminaries

Yongtao Lyu and Yonggang Zheng

1.1 Introduction

Before introducing the "Theory of Elasticity: Basic Concepts and Formulations", first of all, let's have a look at the differences among "Theoretical Mechanics", "Mechanics of Materials" and "Theory of Elasticity". The "Theoretical Mechanics" is based on the assumption of mass point and rigid body, is mainly to investigate the kinematics of the object. Some representative examples in "Theoretical Mechanics" are celestial mechanics, etc. The "Mechanics of Materials" is based on the assumption of continuity, homogeneity, etc., and considers the deformation of the body. In "Mechanics of Materials", simplified geometries (e.g., bar, beam, shell) and simplified forms (e.g., small deformation, plane section) are assumed to solve the simple problems. The "Theory of Elasticity" is an extension of the "Mechanics of Materials" and is to deal with complex scenarios, e.g., statics, dynamics, and it is the basis for all engineering designs and computations, the finite element method, the fracture mechanics and the plastic mechanics, etc.

The "Theory of Elasticity" is concerned with determining the stress, strain and displacement distributions in an elastic solid under the influence of external loadings. It uses the continuum mechanics and formulations to establish the mathematical boundary value problem, and also provides basis for advanced studies in inelastic material behaviors including plasticity, viscoelasticity, etc.

The mathematical theory of elasticity deals with the development of models that describe the response of solid bodies to applied loads. In other words, we typically want to determine the deformation of a solid body and the internal forces within it, given the loads and displacements applied on the boundaries.

Y. Lyu (✉) · Y. Zheng
School of Mechanics and Aerospace Engineering, Dalian University of Technology, Dalian, Liaoning, China
e-mail: Yongtaolu@dlut.edu.cn; zhengyg@dlut.edu.cn

Y. Lyu, Y. Zheng, *Theory of Elasticity*,
https://doi.org/10.1007/978-981-95-6043-1_1

The theory of elasticity can be used to determine the strength of structures such as buildings, bridges, etc. It can also be applied in the fields such as seismology, biomechanics and plant growth to determine the stress, strain, etc. There are three main jobs in the theory of elasticity:

(a) Characterize the deformation and strain;
(b) Characterize the forces both external (loads) and internal (stresses) ones;
(c) Find the relationship between the deformation and forces, i.e., the constitutive law.

The strain and stress are characterized in the same way for different solid materials and the differences between materials arise only through the constitutive law. In this textbook, we consider only the simplest constitutive law corresponding to an isotropic, linear elastic solid material.

Perfectly elastic behavior means that all the deformations are reversible and that the stress depends only on the current state of strain. In other words, there are no history or memory effects and there is no permanent deformation. Materials that retain some memories of their previous stress states are known as viscoelastic materials and the theory of plasticity is used to describe the permanent deformation of material. Most solids will exhibit an elastic behavior under small deformations, but above a critical strain the material will yield and go into the plastically deformation stage and will not return to its original shape. For example, considering a metal paper clip: if you pull the end a little, it will spring back to its original shape, but if you bend the end out, then it will remain in the new shape.

1.2 Basic Concepts in this Textbook

Let's introduce some basic concepts in "Theory of Elasticity" as below:

Elastic body vs. **inelastic body**: an *elastic body* is the one that returns to its original shape after a small deformation. In contrast, an *inelastic body* is the one that does not return to its original shape after a small deformation. For example, a spring is an example of an elastic body, whose deformation is linearly proportionally with the constant of the spring.

Elasticity vs. **plasticity**: *elasticity* refers to the ability of a body to deform in response to the applied forces and to recover to its original shape when the forces are removed. In contrast, *plasticity* is to describe the permanent deformation of the body under exterior loadings.

Linear vs. **nonlinear behaviors**: The *linear behavior* of the deformable body refers to the behavior that the strain is linearly proportional to the stress in the elastic region of the stress-strain curve. In contrast, the *nonlinear behavior* refers to the behavior that the strain is nonlinearly proportional to the stress in the elastic region.

1.3 Basic Assumptions in this Textbook

There are five basic assumptions made in this textbook, i.e., continuum, linear elasticity, isotropy and homogeneity, small deformation and no initial stress. The purpose of making these five basic assumptions is to simplify the mathematical calculations involved in this textbook. It should be noted that in the general engineering problems, these five assumptions may not be satisfied and then most of the equations and concepts presented in this book should be re-considered.

The assumption of continuum is made in the material perspective. The continuum assumption means the matter in the body is continuously distributed and fills in the entire region of the space it occupies. The advantage of this assumption is that because of this assumption, the physical measures such as the strain, stress, etc. are continuous, and thus the continuous functions can be used to describe them. It should be noted that this assumption is always applied in the macro scale and in the micro scale level, the matter is always in discrete state, in which scenario, the finite element method cannot be used and the molecular dynamics method should be used instead.

The assumption of linear elasticity is made in the perspective of the mechanical behavior of the materials. The mechanical behaviors of the materials can be classified into the linear and nonlinear ones. For the nonlinear material, the stress is nonlinearly changing with the strain from the start of the stretch/compression. For the linear material, the stress is linearly changing with the strain at the first stage of the stretch/compression and goes to the nonlinear stage when the material yield point is reached. In this textbook, only the simplest case, i.e., the linear stage in the linear material, is dealt with. Therefore, it is assumed that the constitutive relationship of the material is linear elastic. Because of this assumption, the elastic parameters of the materials are constants which enable us to easily characterize the mechanical behaviors of the material using Young's modulus, shear modulus, elasticity matrix, etc.

The assumption of isotropy and homogeneity is also made in the perspective of the mechanical behavior of material. Depends on the way the deformable body is formed, the material can be classified into the isotropic, the anisotropic, the homogeneous, the heterogeneous, etc. If the deformable body is made of one material and there is the same mechanical behavior in any part and any direction of the object, the material is isotropic and homogeneous. It should be noted that this is the simplest case and only the simplest case is dealt with in this textbook. The advantage of this assumption is that any part of the deformable body can be regarded as the representative volume in the analysis. However, it should be noted that the mechanical behavior of the material can be much more complex. For example, if the deformable body is made of several different materials, the deformable body exhibits the heterogeneous behavior. If the isotropic material is the used as the base material and reinforced with parallel fibers, the material exhibits the transverse isotropic behavior. However, the formulations, especially the finite element formulations, for these materials are out of the scope of this textbook.

The fourth assumption made in this textbook is the small deformation, i.e., the displacement and deformation in the deformable body is much less than its dimension. The assumption of small deformation is made in the perspective of the mechanical behavior of material. The advantage of this assumption is that the dimensional change in the deformable body caused by the deformation can be ignored, and consequently the high order parts in the Taylor series equation can be ignored and subsequent equations can be simplified as the linear elastic equations. It should be noted that the finite element (FE) formulations, especially the geometric equations (strain-displacement equation) presented in this textbook are based on the assumption of small deformation. It should be noted that in the scenario of large deformation, the strain-displacement formulations for the deformable body are completely different and consequently the FE formulations will be different and this is also out of the scope of the present textbook.

The assumption of no initial stress means when the deformable body is cut through in any cross-section of the body, no internal stress will be released and the cutting surfaces in the separated deformable body will not be deformed.

1.4 Mathematical Preliminaries

Similar to other mechanics subjects such as fluid mechanics, heat conduction, and electromagnetics, the theory of elasticity requires the knowledge from mathematics. The theory is formulated in terms of a variety of variables including scalar, vector, and tensor fields, which call for the use of tensor notation along with tensor algebra and calculus. Through the application of principles from the continuum mechanics, the theory is developed as a system of partial differential field equations that are to be solved in a space region coinciding with the body under investigation. Solution techniques used in these field equations commonly employ Fourier methods variational techniques, integral transforms, complex variables, potential theory, finite differences, and finite element method. Therefore, to develop proper formulation methods and solution techniques for elasticity problems, it is necessary to have an appropriate mathematical background. The purpose of this chapter is to provide a background primarily for the mathematical formulation part of this textbook. Additional review of other mathematical topics related to the problem solution technique is provided in later chapters where they are applied.

1.4.1 Definitions for Scalar, Vector, Matrix and Tensor

The theory of elasticity is formulated in terms of many different types of variables that are either specified or sought at spatial points in the body under study. Some of these variables are scalar quantities, representing a single magnitude at each point in space. Common examples of scalar quantities include the material density ρ and temperature T. Other variables of interest are vector quantities that are expressible in terms of components in a two- or three-dimensional coordinate system. Examples of vector variables are the displacement and rotation vectors of material points in the classic continuum body. Formulations within the elasticity theory also require the need for matrix variables, which commonly require more than three components to quantify. Examples of such variables include stress and strain matrices. As shown in the subsequent chapters, a 3D formulation requires nine components (only six are independent) to quantify the stress or strain at a point. In this case, the variable is normally expressed in a matrix format with three rows and three columns. To summarize this discussion, in a 3D Cartesian coordinate system, scalar, vector, and matrix variables can thus be written as follows:

mass density scalar: ρ

displacement vector: $\boldsymbol{u} = u\boldsymbol{e}_1 + v\boldsymbol{e}_2 + w\boldsymbol{e}_3$

stress matrix: $[\boldsymbol{\sigma}] = \begin{bmatrix} \sigma_x & \tau_{xy} & \tau_{xz} \\ \tau_{yx} & \sigma_y & \tau_{yz} \\ \tau_{zx} & \tau_{zy} & \sigma_z \end{bmatrix}$

where $\boldsymbol{e}_1$, $\boldsymbol{e}_2$ and $\boldsymbol{e}_3$ are the usual unit basis vectors in the coordinate directions. Thus, scalars, vectors, and matrices are specified by one, three and nine components, respectively.

The formulation of elasticity problems not only involves these types of variables, but also incorporates additional quantities that require even more components to be characterized. Because of this, most field theories such as elasticity make use of a *tensor formalism* using index notation. This enables efficient representations of all variables and governing equations using a single standardized scheme. The tensor concept will be defined in more details in a later section, but for now we can simply say that scalars, vectors, matrices, and other higher-order variables can all be represented using tensors with various orders.

1.4.2 Index Notation

Index notation is a shorthand scheme whereby variables with a whole set of numbers (elements or components) can be represented by a single symbol with subscripts. For example, a_1, a_2, a_3 can be denoted using a_i, where the index i normally has the range of 1, 2, 3. In a similar fashion, a_{ij} represents nine variables $a_{11}, a_{12}, a_{13}, a_{21}, a_{22}, a_{23}, a_{31}, a_{32}, a_{33}$. Although these representations can be written in many different ways, it is common to use a scheme related to vector and matrix formats such that

$$a_i = \begin{bmatrix} a_1 \\ a_2 \\ a_3 \end{bmatrix}, \quad a_{ij} = \begin{bmatrix} a_{11} & a_{12} & a_{13} \\ a_{21} & a_{22} & a_{23} \\ a_{31} & a_{32} & a_{33} \end{bmatrix} \tag{1.1}$$

In the matrix format, a_{1j} represents the variables in the first row, while a_{i1} indicates the variables in the first column. Other columns and rows are indicated in the similar way, and thus the first index represents the row number, while the second index denotes the column number.

In general, the symbol $a_{ij\ldots k}$ with N distinct indices represents 3^N distinct variables. It is apparent that a_i and a_j represent the same three variables, and likewise a_{ij} and a_{mn} represent the same matrix. Addition, subtraction, multiplication, and equality of these index variables are defined in the normal fashion. For example, addition and subtraction are given as

$$a_i \pm b_i = \begin{bmatrix} a_1 \pm b_1 \\ a_2 \pm b_2 \\ a_3 \pm b_3 \end{bmatrix}$$

$$a_{ij} \pm b_{ij} = \begin{bmatrix} a_{11} \pm b_{11} & a_{12} \pm b_{12} & a_{13} \pm b_{13} \\ a_{21} \pm b_{21} & a_{22} \pm b_{22} & a_{23} \pm b_{23} \\ a_{31} \pm b_{31} & a_{32} \pm b_{32} & a_{33} \pm b_{33} \end{bmatrix} \tag{1.2}$$

and multiplication on scalar variables is specified as

$$\lambda a_i = \begin{bmatrix} \lambda a_1 \\ \lambda a_2 \\ \lambda a_3 \end{bmatrix}, \lambda a_{ij} = \begin{bmatrix} \lambda a_{11} & \lambda a_{12} & \lambda a_{13} \\ \lambda a_{21} & \lambda a_{22} & \lambda a_{23} \\ \lambda a_{31} & \lambda a_{32} & \lambda a_{33} \end{bmatrix} \tag{1.3}$$

The multiplication of two variables with different indices is called *outer multiplication*, and a simple example is given as

$$a_i b_j = \begin{bmatrix} a_1 b_1 & a_1 b_2 & a_1 b_3 \\ a_2 b_1 & a_2 b_2 & a_2 b_3 \\ a_3 b_1 & a_3 b_2 & a_3 b_3 \end{bmatrix} \tag{1.4}$$

The usual commutative, associative, and distributive laws are obeyed, for example:

$$a_i + b_i = b_i + a_i$$

$$a_{ij} b_k = b_k a_{ij}$$

$$a_i + (b_i + c_i) = (a_i + b_i) + c_i$$

$$a_i (b_{jk} c_l) = (a_i b_{jk}) c_l$$

$$a_{ij}(b_k + c_k) = a_{ij} b_k + a_{ij} c_k$$

Note that the simple relations $a_i = b_i$ and $a_{ij} = b_{ij}$ imply that $a_1 = b_1$, $a_2 = b_2$, … and $a_{11} = b_{11}$, $a_{12} = b_{12}$,... However, relations of the form $a_i = b_j$ or $a_{ij} = b_{kl}$ have ambiguous meaning because the distinct indices on each term are not the same, and these types of expressions are to be avoided in this notational scheme. In general, the distinct subscripts on all individual terms in an equation should match.

It is convenient to adopt the convention that if a subscript appears twice in the same term, then summation over that subscript from one to three is implied, for example:

$$a_{ii} = \sum_{i=1}^{3} a_{ii} = a_{11} + a_{22} + a_{33}$$

$$a_{ij}b_j = \sum_{j=1}^{3} a_{ij}b_j = a_{i1}b_1 + a_{i2}b_2 + a_{i3}b_3$$

It is apparent that $a_{ii} = a_{jj} = a_{kk} = \ldots$, and therefore the repeated subscripts or indices are sometimes called *dummy subscripts*. Unspecified indices that are not repeated are called free or distinct subscripts. The summation convention may be suspended by underlining one of the repeated indices or by writing *no sum*. The use of three or more repeated indices in the same term (e.g., a_{iii} or $a_{iij}b_{ij}$) has ambiguous meaning and it should be avoided. On a given symbol, the process of setting two free indices equal is called *contraction*. For example, a_{ii} is obtained from a_{ij} by contraction on i and j. The operation of outer multiplication of two indexed symbols followed by contraction with respect to one index from each symbol generates an *inner multiplication*; for example, $a_{ij}b_{jk}$ is an inner product obtained from the outer product $a_{ij}b_{mk}$ by contraction on indices j and m.

A symbol $a_{ij\ldots m\ldots n\ldots k}$ is said to be symmetric with respect to the index pair mn if

$$a_{ij\ldots m\ldots n\ldots k} = a_{ij\ldots n\ldots m\ldots k} \tag{1.5}$$

while it is *antisymmetric* or *skewsymmetric* if

$$a_{ij\ldots m\ldots n\ldots k} = -a_{ij\ldots n\ldots m\ldots k} \tag{1.6}$$

Note that if $a_{ij\ldots m\ldots n\ldots k}$ is symmetric in mn while $b_{pq\ldots m\ldots n\ldots r}$ is antisymmetric in mn, then the product of them is zero:

$$a_{ij\ldots m\ldots n\ldots k}b_{pq\ldots m\ldots n\ldots r} = 0$$

A second-order tensor may be written as

$$a_{ij} = \frac{1}{2}\left(a_{ij} + a_{ji}\right) + \frac{1}{2}\left(a_{ij} - a_{ji}\right) = a_{(ij)} + \tag{1.7}$$

where, the first term $a_{(ij)} = (a_{ij} + a_{ji})/2$ is symmetric, while the second term $a_{[ij]} = (a_{ij} - a_{ji})/2$ is antisymmetric, and thus an arbitrary tensor a_{ij} can be expressed as the sum of symmetric and antisymmetric parts. Note that if a_{ij} is symmetric, it has only six independent components. On the other hand, if a_{ij} is antisymmetric, its diagonal terms a_{ii} (no sum on i) must be zero, and it has only three independent components.

Example 1.1: Index Notation
Question: The matrix a_{ij} and vector b_i are specified as

$$a_{ij} = \begin{bmatrix} 1 & 2 & 0 \\ 0 & 4 & 3 \\ 2 & 1 & 2 \end{bmatrix}, b_i = \begin{bmatrix} 2 \\ 4 \\ 0 \end{bmatrix}$$

To determine the following quantities: a_{ii}, $a_{ij}a_{ij}$, $a_{ij}a_{jk}$, $a_{ij}b_j$, $a_{ij}b_ib_j$, b_ib_i, b_ib_j, $a_{(ij)}$, $a_{[ij]}$, and indicate whether they are a scalar, vector, or matrix.

Solution: Following the discussion given previously, we can obtain

$$a_{ii} = a_{11} + a_{22} + a_{33} = 7$$

$$a_{ij}a_{ij} = a_{11}a_{11} + a_{12}a_{12} + a_{13}a_{13} + a_{21}a_{21} + a_{22}a_{22} + a_{23}a_{23} + a_{31}a_{31} + a_{32}a_{32} + a_{33}a_{33} = 39$$

$$a_{ij}a_{jk} = a_{i1}a_{1k} + a_{i2}a_{2k} + a_{i3}a_{3k} = \begin{bmatrix} 1 & 10 & 6 \\ 6 & 19 & 18 \\ 6 & 10 & 7 \end{bmatrix}$$

$$a_{ij}b_j = a_{i1}b_1 + a_{i2}b_2 + a_{i3}b_3 = \begin{bmatrix} 10 \\ 16 \\ 8 \end{bmatrix}$$

$$a_{ij}b_ib_j = a_{11}b_1b_1 + a_{12}b_1b_2 + a_{13}b_1b_3 + a_{21}b_2b_1 + \ldots = 84$$

$$b_ib_i = b_1b_1 + b_2b_2 + b_3b_3 = 4 + 16 + 0 = 20$$

$$b_ib_j = \begin{bmatrix} 4 & 8 & 0 \\ 8 & 16 & 0 \\ 0 & 0 & 0 \end{bmatrix}$$

$$a_{(ij)} = \frac{1}{2}\left(a_{ij} + a_{ji}\right) = \frac{1}{2}\begin{bmatrix} 1 & 2 & 0 \\ 0 & 4 & 3 \\ 2 & 1 & 2 \end{bmatrix} + \frac{1}{2}\begin{bmatrix} 1 & 0 & 2 \\ 2 & 4 & 1 \\ 0 & 3 & 2 \end{bmatrix} = \begin{bmatrix} 1 & 1 & 1 \\ 1 & 4 & 2 \\ 1 & 2 & 2 \end{bmatrix}$$

$$a_{[ij]} = \frac{1}{2}\left(a_{ij} - a_{ji}\right) = \frac{1}{2}\begin{bmatrix} 1 & 2 & 0 \\ 0 & 4 & 3 \\ 2 & 1 & 2 \end{bmatrix} - \frac{1}{2}\begin{bmatrix} 1 & 0 & 2 \\ 0 & 4 & 1 \\ 2 & 3 & 2 \end{bmatrix} = \begin{bmatrix} 0 & 1 & -1 \\ -1 & 0 & 1 \\ 1 & -1 & 0 \end{bmatrix}$$

1.4.3 Kronecker Delta and Alternating Symbol

A useful special symbol commonly used in index notational schemes is the *Kronecker delta* defined by

$$\delta_{ij} = \begin{cases} 1, & \text{if } i=j(\text{no sum}) \\ 0, & \text{if } i \neq j \end{cases} = \begin{bmatrix} 1 & 0 & 0 \\ 0 & 1 & 0 \\ 0 & 0 & 1 \end{bmatrix} \tag{1.8}$$

Within usual matrix theory, it is observed that this symbol is simply the unit matrix. Note that the Kronecker delta is a symmetric matrix. Particular useful properties of the Kronecker delta include the following:

$$\delta_{ij} = \delta_{ji}$$

$$\delta_{ii} = 3$$

$$\delta_{ij}a_j = a_i, \delta_{ij}a_i = a_j$$

$$\delta_{ij}a_{jk} = a_{ik}, \delta_{jk}a_{ik} = a_{ij}$$

$$\delta_{ij}a_{ij} = a_{ii}, \delta_{ij}\delta_{ij} = 3$$

Another useful special symbol is the *alternating* or *permutation* symbol defined by

$$\varepsilon_{ijk} = \begin{cases} +1, \text{if } ijk \text{ is an even permutation of } 1,2,3 \\ -1, \text{if } ijk \text{ is an odd permutation of } 1,2,3 \\ 0, \text{otherwise} \end{cases} \tag{1.9}$$

Consequently,

$$\varepsilon_{123} = \varepsilon_{231} = \varepsilon_{312} = 1$$

$$\varepsilon_{321} = \varepsilon_{132} = \varepsilon_{213} = -1$$

$$\varepsilon_{112} = \varepsilon_{131} = \varepsilon_{222} = \ldots = 0$$

Therefore, of the 27 possible terms for the alternating symbol, three of them are equal to +1, three of them are equal to -1, and all others are 0. The alternating symbol is anti-symmetric with respect to any pair of its indices. This particular symbol is useful in determining the determinants and cross products of vectors, and the determinant of an array a_{ij} can be written in two equivalent forms:

$$\det[a_{ij}] = |a_{ij}| = \begin{vmatrix} a_{11} & a_{12} & a_{13} \\ a_{21} & a_{22} & a_{23} \\ a_{31} & a_{32} & a_{33} \end{vmatrix} = \varepsilon_{ijk}a_{1i}a_{2j}a_{3k} = \varepsilon_{ijk}a_{i1}a_{j2}a_{k3} \tag{1.10}$$

where the first index represents the row expansion, and the second is the column expansion. Using the property

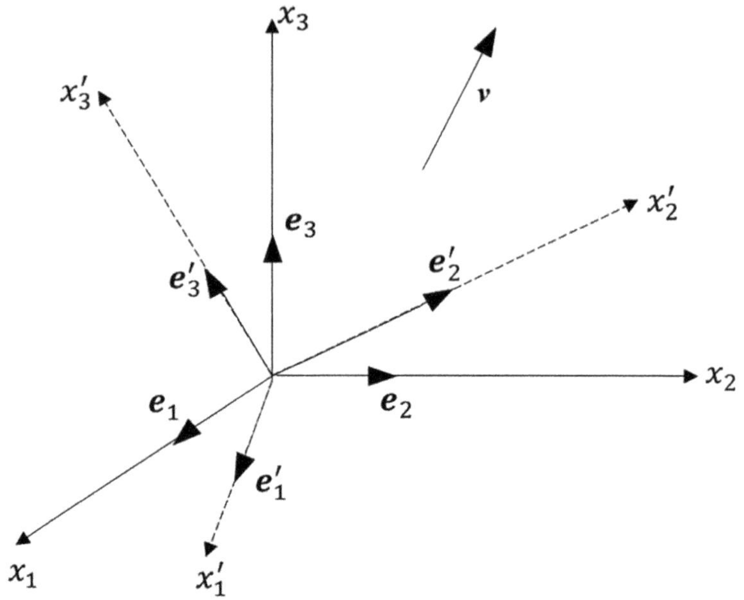

Fig. 1.1 Change of Cartesian coordinate frames

$$\varepsilon_{ijk}\varepsilon_{pqr} = \begin{vmatrix} \delta_{ip} & \delta_{iq} & \delta_{ir} \\ \delta_{jp} & \delta_{jq} & \delta_{jr} \\ \delta_{kp} & \delta_{kq} & \delta_{kr} \end{vmatrix} \tag{1.11}$$

another form of the determinant of a matrix can be written as

$$\det\left[a_{ij}\right] = \frac{1}{6}\varepsilon_{ijk}\varepsilon_{pqr}a_{ip}a_{jq}a_{kr} \tag{1.12}$$

It is convenient and in fact necessary to express the variables and field equations in several different coordinate systems in this textbook. This requires the development of particular transformation rules for scalar, vector, matrix, and higher-order variables. This concept is fundamentally connected to the basic definitions of tensor variables and their related tensor transformations laws. We restrict our discussion to transformations only between Cartesian coordinate systems, and thus the two systems shown in Fig. 1.1 is considered. The two Cartesian frames (x_1, x_2, x_3) and $\left(x'_1, x'_2, x'_3\right)$ differ only by orientation, and the unit basis vectors for each frame are $\{e_i\} = \{e_1, e_2, e_3\}$ and $\left\{e'_i\right\} = \left\{e'_1, e'_2, e'_3\right\}$, respectively.

Let Q_{ij} denote the cosine of the angle between the x'_i-axis and the x_j-axis:

$$Q_{ij} = \cos\left(x'_i, x_j\right) \tag{1.13}$$

Using this definition, the basis vectors in the primed coordinate frame can be easily expressed in terms of those in the unprimed frame by the relations

$$\begin{aligned} \boldsymbol{e}_1' &= Q_{11}\boldsymbol{e}_1 + Q_{12}\boldsymbol{e}_2 + Q_{13}\boldsymbol{e}_3 \\ \boldsymbol{e}_2' &= Q_{21}\boldsymbol{e}_1 + Q_{22}\boldsymbol{e}_2 + Q_{23}\boldsymbol{e}_3 \\ \boldsymbol{e}_3' &= Q_{31}\boldsymbol{e}_1 + Q_{32}\boldsymbol{e}_2 + Q_{33}\boldsymbol{e}_3 \end{aligned} \tag{1.14}$$

or in the index notation

$$\boldsymbol{e}_i' = Q_{ij}\boldsymbol{e}_j \tag{1.15}$$

Likewise, the opposite transformation can be written using the same format as

$$\boldsymbol{e}_i = Q_{ji}\boldsymbol{e}_j' \tag{1.16}$$

Now an arbitrary vector $\boldsymbol{v}$ can be written in either of the two coordinate systems as

$$\begin{aligned} \boldsymbol{v} &= v_1\boldsymbol{e}_1 + v_2\boldsymbol{e}_2 + v_3\boldsymbol{e}_3 = v_i\boldsymbol{e}_i \\ &= v_1'\boldsymbol{e}_1' + v_2'\boldsymbol{e}_2' + v_3'\boldsymbol{e}_3' = v_i'\boldsymbol{e}_i' \end{aligned} \tag{1.17}$$

Substituting Eq. (1.16) into Eq. (1.17) gives

$$\boldsymbol{v} = v_i Q_{ji}\boldsymbol{e}_j' \tag{1.18}$$

From Eq. (1.17), and we find that

$$v_i' = Q_{ij}v_j \tag{1.19}$$

In a similar fashion, substituting Eq. (1.15) into Eq. (1.17) gives

$$v_i = Q_{ji}v_i' \tag{1.20}$$

Equations (1.19) and (1.20) constitute the transformation laws for the Cartesian components of a vector under a change of rectangular Cartesian coordinate frame. It should be understood that under such transformations, the vector is unaltered (retaining original length and orientation), and only its components are changed. Consequently, if we know the components of a vector in one frame, Eq. (1.19) and/or Eq. (1.20) can be used to calculate components in any other frame.

The fact that transformations are being made only between orthogonal coordinate systems places some particular restrictions on the transformation or direction cosine matrix Q_{ij}. These can be determined by using Eqs. (1.19) and (1.20) to get

$$v_i = Q_{ji}v'_j = Q_{ji}Q_{jk}v_k \tag{1.21}$$

From the properties of the Kronecker delta, this expression can be written as

$$\delta_{ik}v_k = Q_{ji}Q_{jk}v_k \text{ or } \left(Q_{ji}Q_{jk} - \delta_{ik}\right)v_k = 0 \tag{1.22}$$

and since this relation is true for all vectors v_k, the expression in parentheses must be zero, giving the result

$$Q_{ji}Q_{jk} = \delta_{ik} \tag{1.23}$$

In a similar fashion, Eqs. (1.19) and (1.20) can be used to eliminate v_i (instead of v'_i) to get

$$Q_{ij}Q_{kj} = \delta_{ik} \tag{1.24}$$

Equations (1.23) and (1.24) comprise the orthogonality conditions that Q_{ij} must satisfy. Taking the determinant of either relation gives another related result:

$$\det\left[Q_{ij}\right] = \pm 1$$

Matrices that satisfy these relations are called orthogonal, and the transformations given by Eqs. (1.19) and (1.20) are therefore referred to as orthogonal transformations.

1.4.4 Cartesian Tensors

Scalars, vectors, matrices, and higher-order quantities can be represented by a general index notational scheme. Using this approach, all quantities may then be referred to as tensors of different orders. The previously presented transformation properties of a vector can be used to establish the general transformation properties of these tensors. Restricting the transformations to those only between Cartesian coordinate systems, the general set of transformation relations for various orders can be written as

$$\begin{cases} a' &= a, \text{zero order (scalar)} \\ a'_i &= Q_{ip}a_p, \text{first order (vector)} \\ a'_{ij} &= Q_{ip}Q_{iq}a_{pq}, \text{second order (matrix)} \\ a'_{ijk} &= Q_{ip}Q_{jq}Q_{kr}a_{pqr}, \text{third order} \\ a'_{ijkl} &= Q_{ip}Q_{jq}Q_{kr}Q_{ls}a_{pqrs}, \text{fourth order} \\ \vdots & \\ a'_{ijk\ldots m} &= Q_{ip}Q_{jq}Q_{kr}\cdots Q_{mt}a_{pqr\ldots t} \text{ general order} \end{cases} \tag{1.25}$$

Note that, according to these definitions, a scalar is a zero-order tensor, a vector is a tensor of order one and a matrix is a tensor of order two. Equation (1.25) then specify the transformation rules for the components of Cartesian tensors of any order under the rotation Q_{ij}. This transformation theory proves to be very valuable in determining the displacement, stress, and strain in different coordinate directions. Some tensors are of a special form in which their components remain the same under all transformations, and these are referred to as isotropic tensors. It can be easily verified that the Kronecker delta δ_{ij} has such a property and is therefore a second-order isotropic tensor. The alternating symbol ε_{ijk} is found to be the third-order isotropic form. The fourth-order case can be expressed in terms of products of Kronecker deltas, and this has important applications in formulating isotropic elastic constitutive relations.

The distinction between the components and the tensor should be noted. Recall that a vector $\boldsymbol{v}$ can be expressed as

$$\begin{aligned} \boldsymbol{v} &= v_1\boldsymbol{e}_1 + v_2\boldsymbol{e}_2 + v_3\boldsymbol{e}_3 = v_i\boldsymbol{e}_i \\ &= v'_1\boldsymbol{e}'_1 + v'_2\boldsymbol{e}'_2 + v'_3\boldsymbol{e}'_3 = v'_i\boldsymbol{e}'_i \end{aligned} \tag{1.26}$$

In a similar fashion, a second-order tensor $\boldsymbol{A}$ can be written as

$$\begin{aligned} \boldsymbol{A} = &\; A_{11}\boldsymbol{e}_1\boldsymbol{e}_1 + A_{12}\boldsymbol{e}_1\boldsymbol{e}_2 + A_{13}\boldsymbol{e}_1\boldsymbol{e}_3 \\ &+ A_{21}\boldsymbol{e}_2\boldsymbol{e}_1 + A_{22}\boldsymbol{e}_2\boldsymbol{e}_2 + A_{23}\boldsymbol{e}_2\boldsymbol{e}_3 \\ &+ A_{31}\boldsymbol{e}_3\boldsymbol{e}_1 + A_{32}\boldsymbol{e}_3\boldsymbol{e}_2 + A_{33}\boldsymbol{e}_3\boldsymbol{e}_3 \\ = &\; A_{ij}\boldsymbol{e}_i\boldsymbol{e}_j = A'_{ij}\boldsymbol{e}'_i\boldsymbol{e}'_j \end{aligned} \tag{1.27}$$

and similar schemes can be used to represent tensors of higher order. The representation used in Eq. (1.27) is commonly called *dyadic notation*, and some authors write the dyadic products $\boldsymbol{e}_i\boldsymbol{e}_j$ using a tensor product notation $\boldsymbol{e}_i \otimes \boldsymbol{e}_j$.

Equations (1.26) and (1.27) indicate that any tensor can be expressed in terms of components in any coordinate system, and it is only the components that change under coordinate transformation. For example, the state of stress at a point in an elastic solid depends on the geometry and applied loadings. As shown later, these stress components are those of a second-order tensor and therefore obey transformation law, i.e. Equation (1.25). Although the components of the stress tensor change with the choice of coordinates, the stress tensor (representing the state of stress) does not.

An important property of a tensor is that if we know its components in one coordinate system, we can find them in any other coordinate frame by using the appropriate transformation law. Because the components of Cartesian tensors are representable by indexed symbols, the operations of equality, addition, subtraction, multiplication, and so forth, are defined in a manner consistent with the index notation procedures previously discussed. The terminology tensor without the adjective Cartesian usually refers to a more general scheme in which the coordinates are not necessarily rectangular Cartesian and the transformations between coordinates are not always orthogonal.

Example 1.2: Example of Transformation
Question: The components of a first-order tensor and a second-order tensor in a particular coordinate frame are given by

$$a_i = \begin{bmatrix} 1 \\ 4 \\ 2 \end{bmatrix}, a_{ij} = \begin{bmatrix} 1 & 0 & 3 \\ 0 & 2 & 2 \\ 3 & 2 & 4 \end{bmatrix}$$

The task is to determine the components of each tensor in a new coordinate system found through a rotation of 60° ($\pi/6$ radians) about the x_3-axis, choosing a counterclockwise rotation when viewing down the negative x_3-axis.

Solution: The original and primed coordinate systems shown in Fig. 1.2 establish the angles between the various axes. The solution starts by determining the rotation matrix for this case:

$$Q_{ij} = \begin{bmatrix} \cos 60^\circ & \cos 30^\circ & \cos 90^\circ \\ \cos 150^\circ & \cos 60^\circ & \cos 90^\circ \\ \cos 90^\circ & \cos 90^\circ & \cos 0^\circ \end{bmatrix} = \begin{bmatrix} 1/2 & \sqrt{3}/2 & 0 \\ -\sqrt{3}/2 & 1/2 & 0 \\ 0 & 0 & 1 \end{bmatrix}$$

The transformation for the vector quantity follows the rules from Eq. (1.25):

$$a_i' = Q_{ij} a_j = \begin{bmatrix} 1/2 & \sqrt{3}/2 & 0 \\ -\sqrt{3}/2 & 1/2 & 0 \\ 0 & 0 & 1 \end{bmatrix} \begin{bmatrix} 1 \\ 4 \\ 2 \end{bmatrix} = \begin{bmatrix} 1/2 + 2\sqrt{3} \\ 2 - \sqrt{3}/2 \\ 2 \end{bmatrix}$$

and the second-order tensor (matrix) is transformed according to Eq. (1.25):

Fig. 1.2 Coordinate transformation

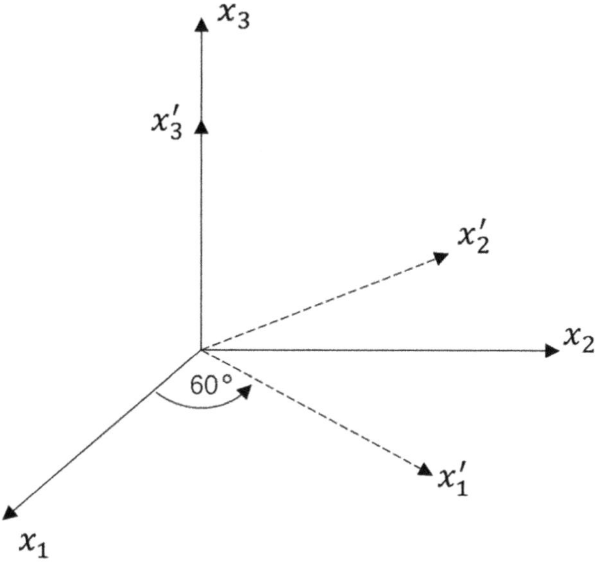

$$a'_{ij} = Q_{ip}Q_{jq}a_{pq} = \begin{bmatrix} 1/2 & \sqrt{3}/2 & 0 \\ -\sqrt{3}/2 & 1/2 & 0 \\ 0 & 0 & 1 \end{bmatrix} \begin{bmatrix} 1 & 0 & 3 \\ 0 & 2 & 2 \\ 3 & 2 & 4 \end{bmatrix} \begin{bmatrix} 1/2 & \sqrt{3}/2 & 0 \\ -\sqrt{3}/2 & 1/2 & 0 \\ 0 & 0 & 1 \end{bmatrix}^T$$

$$= \begin{bmatrix} 7/4 & \sqrt{3}/4 & 3/2+\sqrt{3} \\ \sqrt{3}/4 & 5/4 & 1-3\sqrt{3}/2 \\ 3/2+\sqrt{3} & 1-3\sqrt{3}/2 & 4 \end{bmatrix}$$

where $[\ ]^T$ indicates transpose operation. Although simple transformations can be worked out by hand, for general cases it is more convenient to use a computational scheme to evaluate the necessary matrix multiplications required in the transformation laws Eq. (1.25).

1.4.5 *Principal Values and Directions for Symmetric Second-Order Tensors*

Considering the concept of tensor transformation previously discussed, it should be apparent that there might be particular coordinate systems in which the components of a tensor take on maximum or minimum values. This concept is easily visualized when we consider the components of a vector shown in Fig. 1.1. If we choose a particular coordinate system that has been rotated so that the x_3 axis lies along the direction of the vector, then the vector will have components $\boldsymbol{v} = \{0, 0, |\boldsymbol{v}|\}$. For this case, two of the components have been reduced to zeros, while the remaining component becomes the largest possible (the total magnitude).

This situation is most useful for symmetric second-order tensors that eventually represent the stress and/or strain at a point in an elastic solid. The direction determined by the unit vector n is said to be a principal direction or eigenvector of the symmetric second-order tensor a_{ij} if there exists a parameter λ such that

$$a_{ij}n_j = \lambda n_i \tag{1.28}$$

where λ is called the *principal value* or *eigenvalue of the tensor*. Equation (1.28) can be rewritten as

$$\left(a_{ij} - \lambda\delta_{ij}\right)n_j = 0 \tag{1.29}$$

and this expression is simply a homogeneous system of three linear algebraic equations in the unknowns n_1, n_2, n_3. The system possesses a nontrivial solution if and only if the determinant of its coefficient matrix vanishes, that is:

$$\det\left[a_{ij} - \lambda\delta_{ij}\right] = 0 \tag{1.30}$$

Expanding the determinant produces a cubic equation in terms of λ:

$$\det\left[a_{ij} - \lambda\delta_{ij}\right] = -\lambda^3 + I_1\,\lambda^2 - I_2\,\lambda + I_3 = 0 \tag{1.31}$$

where,

$$\begin{aligned} I_1 &= a_{ii} = a_{11} + a_{22} + a_{33} \\ I_2 &= \frac{1}{2}\left(a_{ii}a_{jj} - a_{ij}a_{ij}\right) = \begin{vmatrix} a_{11} & a_{12} \\ a_{21} & a_{22} \end{vmatrix} + \begin{vmatrix} a_{22} & a_{23} \\ a_{32} & a_{33} \end{vmatrix} + \begin{vmatrix} a_{11} & a_{13} \\ a_{31} & a_{33} \end{vmatrix} \\ I_3 &= \det\left[a_{ij}\right] \end{aligned} \tag{1.32}$$

The scalars I_1, I_2 and I_3 are called the *fundamental invariants* of the tensor a_{ij}, and Eq. (1.31) is known as the *characteristic equation* as indicated by their names, the three invariants do not change their values after coordinate transformation. The roots of the characteristic equation can be used to determine the allowable values for λ, and each of these may be substituted back into Eq. (1.28) to determine the associated principal direction $\boldsymbol{n}$.

Under the condition that the components a_{ij} are real, it can be seen that all three roots λ_1, λ_2, λ_3 of the cubic equation (Eq. (1.31)) must be real. Furthermore, if these roots are distinct, the principal directions associated with each principal value are orthogonal. Thus, we can conclude that every symmetric second-order tensor has at least three mutually perpendicular principal directions and at most three distinct principal values that are the roots of the characteristic equation by denoting the principal directions $\boldsymbol{n}^{(1)}$, $\boldsymbol{n}^{(2)}$, $\boldsymbol{n}^{(3)}$ corresponding to the principal values λ_1, λ_2, λ_3, three possibilities arise:

(a) All three principal values are distinct. Thus, the three corresponding principal directions are unique (except for sense).
(b) Two principal values are equal ($\lambda_1 \neq \lambda_2 = \lambda_3$). The principal direction $\boldsymbol{n}^{(1)}$ is unique (except for sense), and every direction perpendicular to $\boldsymbol{n}^{(1)}$ is a principal direction associated with λ_2, λ_3.
(c) All three principal values are equal. Every direction is principal, and the tensor is isotropic, as per discussion in the previous section.

Therefore, according to what we have presented, it is always possible to identify a right-handed Cartesian coordinate system such that each axis lies along the principal directions of any given symmetric second-order tensor. Such axes are called the *principal axes of the tensor*. For this case, the basis vectors are actually the unit principal directions $\{\boldsymbol{n}^{(1)}, \boldsymbol{n}^{(2)}, \boldsymbol{n}^{(3)}\}$, and it can be seen that with respect to principal axes the tensor reduces to the diagonal form

$$a'_{ij} = \begin{bmatrix} \lambda_1 & 0 & 0 \\ 0 & \lambda_2 & 0 \\ 0 & 0 & \lambda_3 \end{bmatrix} \tag{1.33}$$

Note that the fundamental invariants defined by Eq. (1.32) can be expressed in terms of the principal values as

$$\begin{aligned} I_1 &= \lambda_1 + \lambda_2 + \lambda_3 \\ I_2 &= \lambda_1\lambda_2 + \lambda_2\lambda_3 + \lambda_3\lambda_1 \\ I_3 &= \lambda_1\lambda_2\lambda_3 \end{aligned} \tag{1.34}$$

Example 1.3: Principal Value Problem
Question: Determine the invariants and principal values and directions of the following symmetric second-order tensor:

$$a_{ij} = \begin{bmatrix} 2 & 0 & 0 \\ 0 & 3 & 4 \\ 0 & 4 & -3 \end{bmatrix}$$

The eigenvalues have important extremal properties. If we arbitrarily rank the principal values such that $\lambda_1 > \lambda_2 > \lambda_3$, then λ_1 will be the largest of all possible diagonal elements, while λ_3 will be the smallest of all possible diagonal elements. This theory is applied in elasticity when we seek the largest stress or strain component in an elastic solid.

Solution: The invariants can be calculated as below

$$I_1 = a_{ii} = 2 + 3 - 3 = 2$$

$$I_2 = \begin{vmatrix} 2 & 0 \\ 0 & 3 \end{vmatrix} + \begin{vmatrix} 3 & 4 \\ 4 & -3 \end{vmatrix} + \begin{vmatrix} 2 & 0 \\ 0 & -3 \end{vmatrix} = 6 - 25 - 6 = -25$$

$$I_3 = \begin{vmatrix} 2 & 0 & 0 \\ 0 & 3 & 4 \\ 0 & 4 & -3 \end{vmatrix} = 2(-9 - 16) = -50$$

The characteristic equation then becomes

$$\det\left[a_{ij} - \lambda\delta_{ij}\right] = -\lambda^3 + 2\lambda^2 + 25\lambda - 50 = 0$$

Thus, we have

$$(\lambda - 2)\left(\lambda^2 - 25\right) = 0$$

The solutions for the characteristic equation above are:

$$\lambda_1 = 5, \lambda_2 = 2, \lambda_3 = -5$$

Thus, for this case all principal values are distinct.

For $\lambda_1 = 5$, Eq. (1.28) gives the system

$$\begin{aligned} -3n_1^{(1)} &= 0 \\ -2n_2^{(1)} + 4n_3^{(1)} &= 0 \\ 4n_2^{(1)} - 8n_3^{(1)} &= 0 \end{aligned}$$

which gives a normalized solution $\boldsymbol{n}^{(1)} = \pm(2\boldsymbol{e}_2 + \boldsymbol{e}_3)/\sqrt{5}$. In a similar fashion, the other two principal directions are found to be $\boldsymbol{n}^{(2)} = \pm\boldsymbol{e}_1, \boldsymbol{n}^{(3)} = \pm(\boldsymbol{e}_2 - 2\boldsymbol{e}_3)/\sqrt{5}$. It is easily verified that these directions are mutually orthogonal. Figure 1.3 illustrates their directions with respect to the given coordinate system, and this establishes the right-handed principal coordinate axes $\left(x_1', x_2', x_3'\right)$. For this case, the transformation matrix Q_{ij} defined by Eq. (1.13) becomes

$$Q_{ij} = \begin{bmatrix} 0 & 2/\sqrt{5} & 1/\sqrt{5} \\ 1 & 0 & 0 \\ 0 & 1/\sqrt{5} & -2/\sqrt{5} \end{bmatrix}$$

Notice the eigenvectors actually form the rows of the Q-matrix.

Using this in the transformation law, the components of the given second-order tensor become

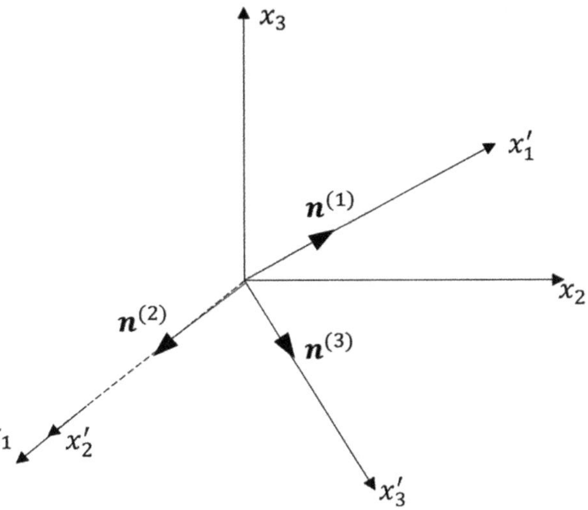

Fig. 1.3 Principal axes for the problem given in Example 1.3

$$a'_{ij} = \begin{bmatrix} 5 & 0 & 0 \\ 0 & 2 & 0 \\ 0 & 0 & -5 \end{bmatrix}$$

This result then validates the general theory given by Eq. (1.33) indicating that the tensor should take on the diagonal form with the principal values as the elements.

Only simple second-order tensors lead to a characteristic equation that is factorable, thus allowing it to be solved by hand calculation. Most other cases normally develop a general cubic equation and a more complicated system to be solved for the principal directions.

1.4.6 Vector, Matrix, and Tensor Algebra

The theory of elasticity requires the use of many standard algebraic operations among vector, matrix, and tensor variables. These operations include dot and cross products of vectors and numerous matrix/tensor products. All these operations can be expressed efficiently using compact tensor index notation. First, some particular vector products are considered. Given two vectors $\boldsymbol{a}$ and $\boldsymbol{b}$, with Cartesian components a_i and b_i, the scalar or dot product is defined as

$$\boldsymbol{a} \cdot \boldsymbol{b} = a_1 b_1 + a_2 b_2 + a_3 b_3 = a_i b_i \tag{1.35}$$

Because all indices in this expression are repeated, the quantity should be a scalar, that is, a tensor of order zero. The magnitude of a vector can then be expressed as

$$|\boldsymbol{a}| = (\boldsymbol{a}\cdot\boldsymbol{a})^{1/2} = (a_i a_i)^{1/2} \tag{1.36}$$

The vector or cross product between two vectors $\boldsymbol{a}$ and $\boldsymbol{b}$ can be written as

$$\boldsymbol{a}\times\boldsymbol{b} = \begin{vmatrix} \boldsymbol{e}_1 & \boldsymbol{e}_2 & \boldsymbol{e}_3 \\ a_1 & a_2 & a_3 \\ b_1 & b_2 & b_3 \end{vmatrix} = \varepsilon_{ijk} a_j b_k \boldsymbol{e}_i \tag{1.37}$$

where $\boldsymbol{e}_i$ are the unit basis vectors for the coordinate system. Note that the cross product gives a vector resultant whose components are $\varepsilon_{jk} a_j b_k$. Another common vector product is the scalar triple product defined by

$$\boldsymbol{a}\cdot\boldsymbol{b}\times\boldsymbol{c} = \begin{vmatrix} a_1 & a_2 & a_3 \\ b_1 & b_2 & b_3 \\ c_1 & c_2 & c_3 \end{vmatrix} = \varepsilon_{ijk} a_i b_j c_k \tag{1.38}$$

Next, considering some common matrix products, i.e., using the usual direct notation for matrices and vectors, common products between a matrix $\boldsymbol{A} = [\boldsymbol{A}]$ and a vector $\boldsymbol{a}$ can be written as

$$\begin{aligned} \boldsymbol{A}\boldsymbol{a} &= [\boldsymbol{A}]\{\boldsymbol{a}\} = A_{ij} a_j = a_j A_{ij} \\ \boldsymbol{a}^T\boldsymbol{A} &= \{\boldsymbol{a}\}^T[\boldsymbol{A}] = a_i A_{ij} = A_{ij} a_j \end{aligned} \tag{1.39}$$

where $\boldsymbol{a}^T$ denotes the transpose operation of $\boldsymbol{a}$, and for a vector quantity, this simply changes the column matrix (3×1) into a row matrix (1×3). Note that each of these products results in a vector resultant. These types of expressions generally involve various inner products within the index notational scheme, and as noted, once the summation index is properly specified, the order of listing the product terms does not change the result. We will encounter several different combinations of products between two matrices $\boldsymbol{A}$ and $\boldsymbol{B}$:

$$\begin{aligned} \boldsymbol{A}\boldsymbol{B} &= [\boldsymbol{A}][\boldsymbol{B}] = A_{ij} B_{jk} \\ \boldsymbol{A}\boldsymbol{B}^T &= A_{ij} B_{kj} \\ \boldsymbol{A}^T\boldsymbol{B} &= A_{ji} B_{jk} \\ \mathrm{tr}(\boldsymbol{A}\boldsymbol{B}) &= A_{ij} B_{ji} \\ \mathrm{tr}\left(\boldsymbol{A}\boldsymbol{B}^T\right) = \mathrm{tr}\left(\boldsymbol{A}^T\boldsymbol{B}\right) &= A_{ij} B_{ij} \end{aligned} \tag{1.40}$$

where $\boldsymbol{A}^T$ indicates the transpose operation of $\boldsymbol{A}$ and $tr\boldsymbol{A}$ is the trace of the matrix defined by

$$\begin{aligned} A_{ij}^{T} &= A_{\tilde{ji}} \\ trA &= A_{ii} = A_{11} + A_{22} + A_{33} \end{aligned} \tag{1.41}$$

Similar to vector products, once the summation index is properly specified, the results in Eq. (1.40) do not depend on the order of listing the product terms. Please be noted that this does not imply that $\boldsymbol{AB} = \boldsymbol{BA}$, which is certainly not true.

1.4.7 Calculus of Cartesian Tensors

Most variables within the theory of elasticity are the field variables, that is, functions depending on the spatial coordinates are used to formulate the problem in this textbook. For time-dependent problems, these variables could also have temporal variations. Thus, scalar, vector, matrix, and general tensor variables are functions of the spatial coordinates (x_1, x_2, x_3). Because many elasticity equations involve the differential and integral operations, it is necessary to have an understanding of the calculus of Cartesian tensor fields. The field concept for tensor components can be expressed as

$$\begin{aligned} a &= a(x_1, x_2, x_3) = a(x_i) = a(\boldsymbol{x}) \\ a_i &= a_i(x_1, x_2, x_3) = a_i(x_i) = a_i(\boldsymbol{x}) \\ a_{ij} &= a_{ij}(x_1, x_2, x_3) = a_{ij}(x_i) = a_{ij}(\boldsymbol{x}) \\ &\vdots \end{aligned} \tag{1.42}$$

It is convenient to introduce the comma notation for partial differentiation:

$$a_{,i} = \frac{\partial}{\partial x_i} a; \, a_{i,j} = \frac{\partial}{\partial x_j} a_i; \, a_{ij,k} = \frac{\partial}{\partial x_k} a_{ij}; \, \ldots$$

It can be seen that if the differentiation index is distinct, the order of the tensor is increased by one. For example, the derivative operation on a vector a_i produces a second-order tensor or matrix given by

$$a_{i,j} = \begin{bmatrix} \frac{\partial a_1}{\partial x_1} & \frac{\partial a_1}{\partial x_2} & \frac{\partial a_1}{\partial x_3} \\ \frac{\partial a_2}{\partial x_1} & \frac{\partial a_2}{\partial x_2} & \frac{\partial a_2}{\partial x_3} \\ \frac{\partial a_3}{\partial x_1} & \frac{\partial a_3}{\partial x_2} & \frac{\partial a_3}{\partial x_3} \end{bmatrix} \tag{1.43}$$

Using Cartesian coordinates (x, y, z), the directional derivative of a scalar field function f with respect to a direction s can be worked out as:

$$\frac{df}{ds} = \frac{\partial f}{\partial x}\frac{dx}{ds} + \frac{\partial f}{\partial y}\frac{dy}{ds} + \frac{\partial f}{\partial z}\frac{dz}{ds} \tag{1.44}$$

Note that the unit vector in the direction of s can be written as

$$\boldsymbol{n} = \frac{dx}{ds}\boldsymbol{e}_1 + \frac{dy}{ds}\boldsymbol{e}_2 + \frac{dz}{ds}\boldsymbol{e}_3 \tag{1.45}$$

Therefore, the directional derivative can be expressed as the following scalar product:

$$\frac{df}{ds} = \boldsymbol{n} \cdot \nabla f \tag{1.46}$$

where ∇f is called the *gradient of the scalar function f* and is defined as

$$\nabla f = \text{grad} f = \boldsymbol{e}_1 \frac{\partial f}{\partial x} + \boldsymbol{e}_2 \frac{\partial f}{\partial y} + \boldsymbol{e}_3 \frac{\partial f}{\partial z} \tag{1.47}$$

and the symbolic vector operator ∇ is called the *del operator*

$$\nabla = \boldsymbol{e}_1 \frac{\partial}{\partial x} + \boldsymbol{e}_2 \frac{\partial}{\partial y} + \boldsymbol{e}_3 \frac{\partial}{\partial z} \tag{1.48}$$

These and other useful operations can be expressed in Cartesian tensor notation. Given the scalar field ϕ and vector field $\boldsymbol{u}$, the following common differential operations can be written in the index notation:

$$\begin{cases} \text{Gradient of a Scalar}: \nabla\phi = \phi_{,i}\boldsymbol{e}_i \\ \text{Gradient of a Vector}: \nabla\boldsymbol{u} = u_{i,j}\boldsymbol{e}_i\boldsymbol{e}_j \\ \text{Laplacian of a Scalar}: \nabla^2\phi = \nabla\cdot\nabla\phi = \phi_{,ii} \\ \text{Divergence of a Vector}: \nabla\cdot\boldsymbol{u} = u_{i,i} \\ \text{Curl of a Vector}: \nabla\times\boldsymbol{u} = \varepsilon_{ijk}u_{k,j}\boldsymbol{e}_i \\ \text{Laplacian of a Vector}: \nabla^2\boldsymbol{u} = u_{i,kk}\boldsymbol{e}_i \end{cases} \tag{1.49}$$

If ϕ and ψ are scalar fields and $\boldsymbol{u}$ and $\boldsymbol{v}$ are vector fields, several useful identities exist:

$$\begin{cases} \nabla(\phi\psi) = (\nabla\phi)\psi + \phi(\nabla\psi) \\ \nabla^2(\phi\psi) = (\nabla^2\phi)\psi + \phi(\nabla^2\psi) + 2\nabla\phi\cdot\nabla\psi \\ \nabla\cdot(\phi\boldsymbol{u}) = \nabla\phi\cdot\boldsymbol{u} + \phi(\nabla\cdot\boldsymbol{u}) \\ \nabla\times(\phi\boldsymbol{u}) = \nabla\phi\times\boldsymbol{u} + \phi(\nabla\times\boldsymbol{u}) \\ \nabla\cdot(\boldsymbol{u}\times\boldsymbol{v}) = \boldsymbol{v}\cdot(\nabla\times\boldsymbol{u}) - \boldsymbol{u}\cdot(\nabla\times\boldsymbol{v}) \\ \nabla\times\nabla\phi = 0 \\ \nabla\cdot\nabla\phi = \nabla^2\phi \\ \nabla\cdot\nabla\times\boldsymbol{u} = 0 \\ \nabla\times(\nabla\times\boldsymbol{u}) = \nabla(\nabla\cdot\boldsymbol{u}) - \nabla^2\boldsymbol{u} \\ \boldsymbol{u}\times(\nabla\times\boldsymbol{u}) = \frac{1}{2}\nabla(\boldsymbol{u}\cdot\boldsymbol{u}) - \boldsymbol{u}\cdot\nabla\boldsymbol{u} \end{cases} \tag{1.50}$$

Each of these identities can be easily justified using index notation from Eq. (1.49).

Example 1.4: Scalar and Vector Field
Question: Scalar and vector field functions are given by $\phi = x^2 - y^2$ and $=2x\boldsymbol{e}_1 + 3yz\boldsymbol{e}_2 + xy\boldsymbol{e}_3$, respectively. The tasks are to work out the values of the following expressions, $\nabla\phi$, $\nabla^2\phi$, $\nabla \cdot \boldsymbol{u}$, $\nabla\,\boldsymbol{u}$, $\nabla \times \boldsymbol{u}$.

Solution: Using the basic relations in Eq. (1.49)

$$\nabla\phi = 2x\boldsymbol{e}_1 - 2y\boldsymbol{e}_2$$

$$\nabla^2\phi = 2 - 2 = 0$$

$$\nabla\cdot\boldsymbol{u} = 2 + 3z + 0 = 2 + 3z$$

$$\nabla\boldsymbol{u} = u_{i,j} = \begin{bmatrix} 2 & 0 & 0 \\ 0 & 3z & 3y \\ y & x & 0 \end{bmatrix}$$

$$\nabla\times\boldsymbol{u} = \begin{vmatrix} \boldsymbol{e}_1 & \boldsymbol{e}_2 & \boldsymbol{e}_3 \\ \partial/\partial x & \partial/\partial y & \partial/\partial z \\ 2x & 3yz & xy \end{vmatrix} = (x - 3y)\boldsymbol{e}_1 - y\boldsymbol{e}_2$$

Using numerical methods, some of these variables can be conveniently computed and plotted in order to visualize the nature of the field distribution. For example, contours of $\phi=$ constant can be easily plotted using Matlab software, and vector distributions of $\nabla\phi$ can be shown as plots of vectors properly scaled in magnitude and orientation. Fig. 1.4 shows these two types of plots, and it is observed that the vector field $\nabla\phi$ is orthogonal to the ϕ-contours, a result that is true in general for all scalar fields. Numerically generated plots such as these will prove to be useful in understanding displacement, strain, and/or stress distributions found in the solution to a variety of problems in elasticity.

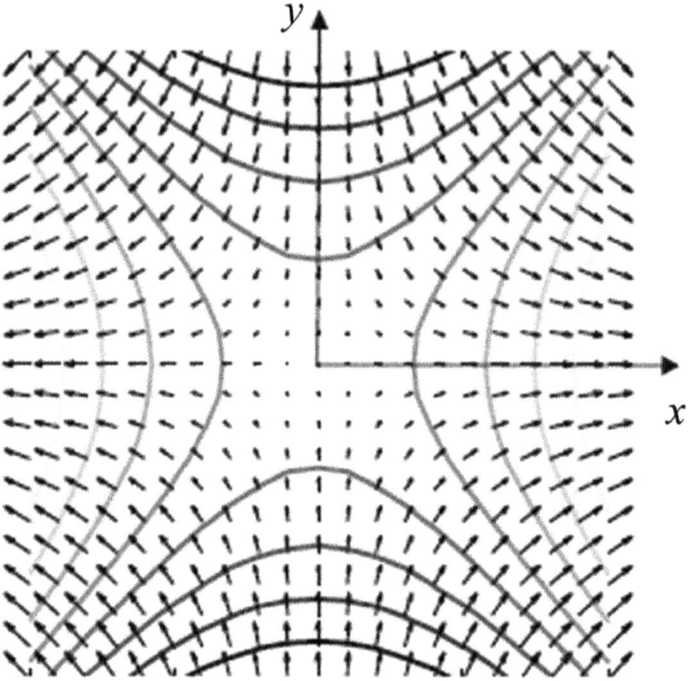

Fig. 1.4 Contours of ϕ = constant and vector distributions of $\nabla\phi$

1.4.8 Vectors

We denote vector quantities using boldface letters (e.g., **A**). The basis vectors in Cartesian coordinates are denoted as $\boldsymbol{e}_1$, $\boldsymbol{e}_2$ and $\boldsymbol{e}_3$ in the x_1, x_2 and x_3 directions, respectively. We prefer to use numerical subscripts for the vector components and basis vectors, since the equations of elasticity can be developed in a very compact form using index notation and Cartesian tensors. Thus, in Cartesian coordinates, the vector $\boldsymbol{x}$ can be written as

$$\boldsymbol{x} = x_1\boldsymbol{e}_1 + x_2\boldsymbol{e}_2 + x_3\boldsymbol{e}_3 \tag{1.51}$$

where x_1, x_2 and x_3 are the Cartesian components of $\boldsymbol{x}$.

Note that the basis vectors are unit vectors (vectors with unit length), so that

$$|\boldsymbol{e}_1| = |\boldsymbol{e}_2| = |\boldsymbol{e}_3| = 1 \text{ or } |\boldsymbol{e}_i| = 1, i = 1, 2, 3 \tag{1.52}$$

The basis vectors are also mutually perpendicular:

$$\boldsymbol{e}_1 \cdot \boldsymbol{e}_2 = \boldsymbol{e}_2 \cdot \boldsymbol{e}_3 = \boldsymbol{e}_3 \cdot \boldsymbol{e}_1 = 0 \tag{1.53}$$

In general,

$$\boldsymbol{e}_i \cdot \boldsymbol{e}_j = \delta_{ij} \tag{1.54}$$

where δ_{ij} is the Kronecker delta defined as

$$\delta_{ij} = \begin{cases} 1, & i = j \\ 0, & i \neq j \end{cases} \tag{1.55}$$

Thus, the three basis vectors form an orthonormal basis (a basis whose basis vectors are mutually orthogonal unit vectors).

We can also write $\boldsymbol{x}$ using the summation convention, where repeated indices imply the summation over the range (1, 2, 3 in three dimensions), and simply write

$$\boldsymbol{x} = x_i \boldsymbol{e}_i \tag{1.56}$$

In Eq. (1.56), the summation over the range 1, 2, 3 is implied, since the dummy index i appears twice. Using the summation convention, the Kronecker delta can be rewritten as

$$\delta_{ii} = \begin{cases} \delta_{11} + \delta_{22} = 2 \text{ in 2D} \\ \delta_{11} + \delta_{22} + \delta_{33} = 3 \text{ in 3D} \end{cases} \tag{1.57}$$

Since the Kronecker delta is the index notation form of the identity matrix $\mathbf{I}$, δ_{ii} is the trace of $\mathbf{I}$ (the sum of the diagonal entries). On the other hand, the unit basis vectors form a right-handed orthogonal triad,

$$\begin{aligned} \boldsymbol{e}_1 \times \boldsymbol{e}_2 = \boldsymbol{e}_3, \boldsymbol{e}_2 \times \boldsymbol{e}_3 = \boldsymbol{e}_1, \boldsymbol{e}_3 \times \boldsymbol{e}_1 = \boldsymbol{e}_2 \\ \boldsymbol{e}_2 \times \boldsymbol{e}_1 = -\boldsymbol{e}_3, \boldsymbol{e}_3 \times \boldsymbol{e}_2 = -\boldsymbol{e}_1, \boldsymbol{e}_1 \times \boldsymbol{e}_3 = -\boldsymbol{e}_2 \end{aligned} \tag{1.58}$$

These relations can be summarized using the single equation

$$\boldsymbol{e}_i \times \boldsymbol{e}_j = e_{ijk} \boldsymbol{e}_k \tag{1.59}$$

where the dummy index k is summed, and e_{ijk} is the *alternating symbol* (or *permutation symbol*) which is defined as

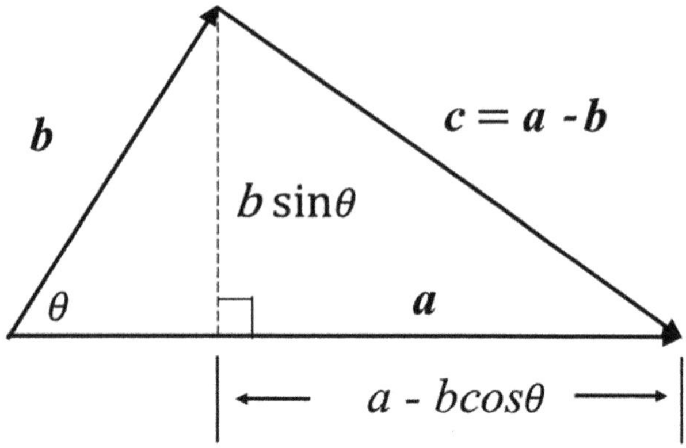

Fig. 1.5 Relation among there vectors

$$e_{ijk} = \begin{cases} +1, & \text{if the subscripts form an even permutation of 123} \\ -1, & \text{if the subscripts form an odd permutation of 123} \\ 0, & \text{otherwise (e.g., if two subscripts are equal).} \end{cases} \quad (1.60)$$

For example,

$$e_{123} = e_{231} = e_{312} = 1$$
$$e_{213} = e_{321} = e_{132} = -1$$
$$e_{113} = e_{133} = e_{221} = 0$$

Considering two vectors $\boldsymbol{a}$ and $\boldsymbol{b}$ (Fig. 1.5) given by

$$\boldsymbol{a} = a_i \boldsymbol{e}_i, \boldsymbol{b} = b_i \boldsymbol{e}_i \quad (1.61)$$

Since the index i is a dummy index, any symbol can be used. Thus, we can also write

$$\boldsymbol{b} = b_j \boldsymbol{e}_j \quad (1.62)$$

The scalar (or dot) product of $\mathbf{a}$ and $\boldsymbol{b}$ is

$$\boldsymbol{a} \cdot \boldsymbol{b} = (a_i \boldsymbol{e}_i) \cdot \left(b_j \boldsymbol{e}_j\right) = a_i b_j \left(\boldsymbol{e}_i \cdot \boldsymbol{e}_j\right) = a_i b_j \delta_{ij} = a_i b_i \quad (1.63)$$

or, in expanded form,

$$\boldsymbol{a} \cdot \boldsymbol{b} = a_1 b_1 + a_2 b_2 + a_3 b_3 \quad (1.64)$$

We define a third vector $\boldsymbol{c}$ as

$$c = a - b \tag{1.65}$$

as shown in Fig. 1.5, from which we can obtain

$$(a - b\cos\theta)^2 + (b\sin\theta)^2 = c^2 \tag{1.66}$$

where a, b and c are the lengths of $\boldsymbol{a}$, $\boldsymbol{b}$ and $\boldsymbol{c}$, respectively. The length (or magnitude) of the vector $\boldsymbol{a}$ is

$$a = |\boldsymbol{a}| = \sqrt{\boldsymbol{a}\cdot\boldsymbol{a}} = \sqrt{a_i a_i} = \sqrt{a_1^2 + a_2^2 + a_3^2} \tag{1.67}$$

Equation (1.66) can be expanded to yield

$$a^2 - 2ab\cos\theta + b^2\cos^2\theta + b^2\sin^2\theta = c^2 \tag{1.68}$$

or

$$c^2 = a^2 + b^2 - 2ab\cos\theta$$

which is the law of cosines. We can further expand the law of cosines in terms of components to obtain

$$(a_1 - b_1)^2 + (a_2 - b_2)^2 + (a_3 - b_3)^2 = a_1^2 + a_2^2 + a_3^2 + b_1^2 + b_2^2 + b_3^2 - 2ab\cos\theta$$

or

$$a_1 b_1 + a_2 b_2 + a_3 b_3 = ab\cos\theta$$

Thus, the dot product of two vectors can alternatively be expressed as

$$\boldsymbol{a}\cdot\boldsymbol{b} = |\boldsymbol{a}||\boldsymbol{b}|\cos\theta = ab\cos\theta \tag{1.69}$$

where θ is the angle between the two vectors.

The vector (or cross) product of two vectors $\boldsymbol{a}$ and $\boldsymbol{b}$ is

$$\boldsymbol{a}\times\boldsymbol{b} = a_i\boldsymbol{e}_i \times b_j\boldsymbol{e}_j = a_i b_j \boldsymbol{e}_i \times \boldsymbol{e}_j = a_i b_j e_{ijk}\boldsymbol{e}_k \tag{1.70}$$

where the final result follows the laws defined in Eq. (1.59). This result is expanded by the summation over i, j, k to obtain

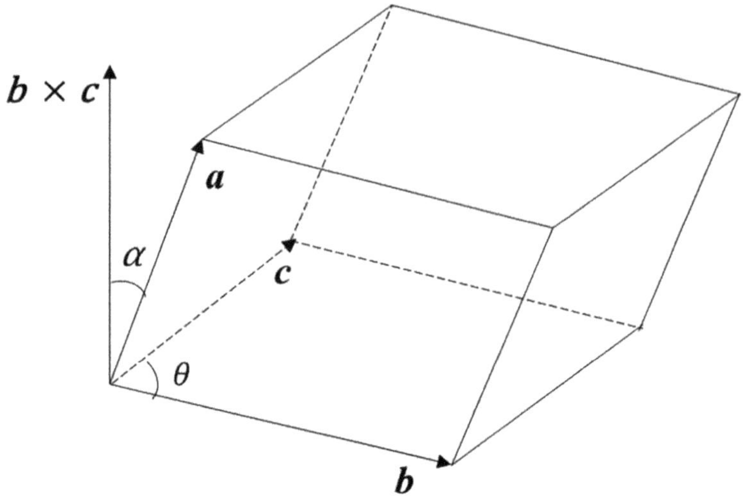

Fig. 1.6 Parallelepiped and triple scalar product

$$\begin{aligned}\boldsymbol{a}\times\boldsymbol{b} &= (a_2b_3-a_3b_2)\boldsymbol{e}_1+(a_3b_1-a_1b_3)\boldsymbol{e}_2+(a_1b_2-a_2b_1)\boldsymbol{e}_3 \\ &= \begin{vmatrix}\boldsymbol{e}_1 & \boldsymbol{e}_2 & \boldsymbol{e}_3\\ a_1 & a_2 & a_3\\ b_1 & b_2 & b_3\end{vmatrix}\end{aligned} \tag{1.71}$$

It is also shown in elementary vector analysis that

$$\boldsymbol{a}\times\boldsymbol{b}=|\boldsymbol{a}||\boldsymbol{b}|(\sin\theta)\boldsymbol{e}_n \tag{1.72}$$

where $\boldsymbol{e}_n$ is the unit vector which is perpendicular to the plane formed by $\boldsymbol{a}$ and $\boldsymbol{b}$ and lies in the direction indicated by the right-hand rule if $\boldsymbol{a}$ is rotated into $\boldsymbol{b}$.

For three vectors $\boldsymbol{a}$, $\boldsymbol{b}$ and $\boldsymbol{c}$, considering the triple scalar product $\boldsymbol{a}\cdot(\boldsymbol{b}\times\boldsymbol{c})$, where **c** is not the same vector defined in Eq. (1.65), this product represents the volume of the parallelepiped having $\boldsymbol{a}$, $\boldsymbol{b}$ and $\boldsymbol{c}$ as edges (Fig. 1.6), since

$$|\boldsymbol{b}\times\boldsymbol{c}|=|\boldsymbol{b}||\boldsymbol{c}|\sin\theta \tag{1.73}$$

is the area of the parallelogram with sides b and c, and $|\boldsymbol{a}|\cos\alpha$ is the height of the parallelepiped. Thus,

$$|\boldsymbol{a}\cdot(\boldsymbol{b}\times\boldsymbol{c})|=|\boldsymbol{a}||\boldsymbol{b}||\boldsymbol{c}|\sin\theta\cos\alpha=\text{height}\times\text{area}=\text{volume}$$

From Eq. (1.71), we can obtain

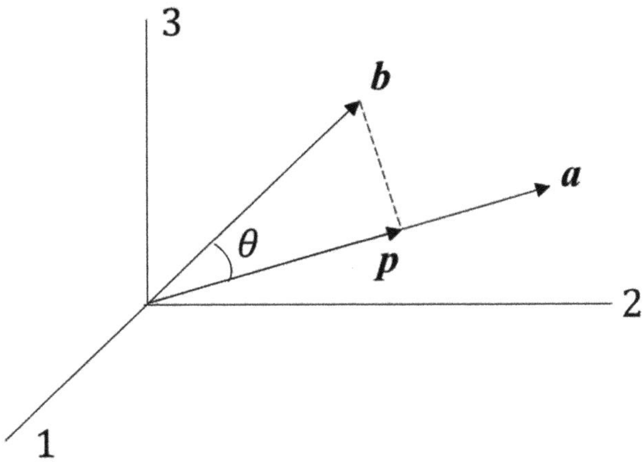

Fig. 1.7 Projection of vector $\boldsymbol{b}$ onto another vector $\boldsymbol{a}$

$$\boldsymbol{a}\cdot(\boldsymbol{b}\times\boldsymbol{c})=\begin{vmatrix} a_1 & a_2 & a_3 \\ b_1 & b_2 & b_3 \\ c_1 & c_2 & c_3 \end{vmatrix} \tag{1.74}$$

The volume property of the triple scalar product implies that

$$\boldsymbol{a}\cdot(\boldsymbol{b}\times\boldsymbol{c})=\boldsymbol{b}\cdot(\boldsymbol{c}\times\boldsymbol{a})=\boldsymbol{c}\cdot(\boldsymbol{a}\times\boldsymbol{b}) \tag{1.75}$$

Therefore, in the triple scalar product, the dot and the cross can be interchanged without affecting the result.

Two vectors $\boldsymbol{a} = (a_1, a_2, a_3)$ and $\boldsymbol{b} = (b_1, b_2, b_3)$, as shown in Fig. 1.7 are considered. Let $\boldsymbol{p}$ be the vector obtained by projecting $\boldsymbol{b}$ onto $\boldsymbol{a}$. The scalar projection of $\boldsymbol{b}$ onto $\boldsymbol{a}$ is $|\boldsymbol{b}|\ \cos\theta$, where θ is the angle between $\boldsymbol{b}$ and $\boldsymbol{a}$. Thus, the vector projection of $\boldsymbol{b}$ onto $\boldsymbol{a}$ is

$$\boldsymbol{p}=\left(|\boldsymbol{b}|\cos\theta\right)\frac{\boldsymbol{a}}{|\boldsymbol{a}|} \tag{1.76}$$

where the fraction in this expression is the unit vector in the direction of $\boldsymbol{a}$. Since

$$\boldsymbol{a}\cdot\boldsymbol{b}=|\boldsymbol{a}||\boldsymbol{b}|\cos\theta \tag{1.77}$$

Equation (1.76) becomes

$$\boldsymbol{p}=\left(\boldsymbol{b}\cdot\frac{\boldsymbol{a}}{|\boldsymbol{a}|}\right)\frac{\boldsymbol{a}}{|\boldsymbol{a}|} \tag{1.78}$$

1.4.9 Change of Basis

On many occasions in engineering applications, including elasticity, the need arises to transform vectors and matrices from one coordinate system to another. Considering the vector $\boldsymbol{v}$ given by

$$\boldsymbol{v} = v_1\boldsymbol{e}_1 + v_2\boldsymbol{e}_2 + v_3\boldsymbol{e}_3 = v_i\boldsymbol{e}_i \tag{1.79}$$

where $\boldsymbol{e}_i$ are the basis vectors, v_i are the components of $\boldsymbol{v}$, and the summation convention was used.

Since bases are not unique, we can express $\boldsymbol{v}$ in two different orthonormal bases:

$$\boldsymbol{v} = \sum_{i=1}^{3} v_i\boldsymbol{e}_i = \sum_{i=1}^{3} \overline{v}_i\overline{\boldsymbol{e}}_i \tag{1.80}$$

where v_i are the components of $\boldsymbol{v}$ in the unbarred coordinate system, and $\overline{v}_i$ are the components in the barred system (Fig. 1.8). If we take the dot product on both sides of Eq. (1.80) with $\boldsymbol{e}_j$, we can obtain

$$\sum_{i=1}^{3} v_i\boldsymbol{e}_i \cdot \boldsymbol{e}_j = \sum_{i=1}^{3} \overline{v}_i\overline{\boldsymbol{e}}_i \cdot \boldsymbol{e}_j \tag{1.82}$$

where $\boldsymbol{e}_i \cdot \boldsymbol{e}_j = \delta_{ij}$, and we define the 3×3 matrix $\boldsymbol{R}$ as

$$R_{ij} = \overline{\boldsymbol{e}}_i \cdot \boldsymbol{e}_j \tag{1.82}$$

Thus, from Eq. (1.81), we can obtain

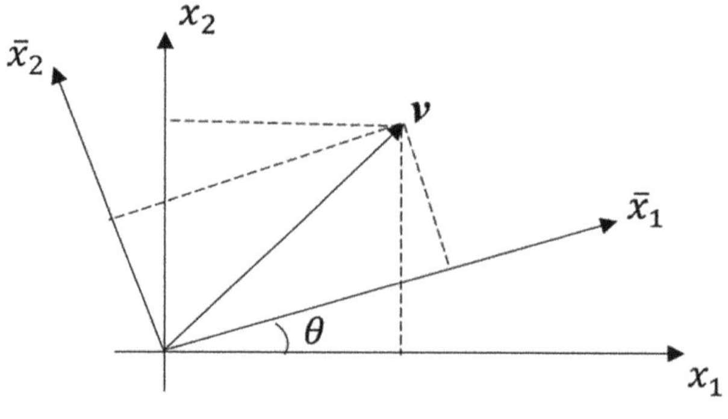

Fig. 1.8 Change of the basis

$$v_j = \sum_{i=1}^{3} R_{ij}\bar{v}_i = \sum_{i=1}^{3} R^T_{ji}\bar{v}_i \tag{1.83}$$

Since the matrix product

$$\boldsymbol{C} = \boldsymbol{A}\boldsymbol{B} \tag{1.84}$$

can be written using subscript notation as $C_{ij} = \sum_{k=1}^{3} A_{ik}B_{kj}$. Equation (1.83) is equivalent to the matrix product

$$\boldsymbol{v} = \boldsymbol{R}^T\bar{\boldsymbol{v}} \tag{1.85}$$

Similarly, if we take the dot product of Eq. (1.80) with $\boldsymbol{e}_j$, we obtain

$$\sum_{i=1}^{3} v_i\boldsymbol{e}_i \cdot \bar{\boldsymbol{e}}_j = \sum_{i=1}^{3} \bar{v}_i\bar{\boldsymbol{e}}_i \cdot \bar{\boldsymbol{e}}_j \tag{1.86}$$

where $\bar{\boldsymbol{e}}_i \cdot \bar{\boldsymbol{e}}_j = \delta_{ij}$, and $\boldsymbol{e}_i \cdot \bar{\boldsymbol{e}}_j = R_{ji}$. Therefore,

$$\bar{v}_j = \sum_{i=1}^{3} R_{ji}v_i \quad \text{or} \quad \bar{\boldsymbol{v}} = \boldsymbol{R}\boldsymbol{v} \quad \text{or} \quad \boldsymbol{v} = \boldsymbol{R}^{-1}\bar{\boldsymbol{v}} \tag{1.87}$$

A comparison of Eq. (1.85) with Eq. (1.87) yields

$$\boldsymbol{R}^{-1} = \boldsymbol{R}^T \quad \text{or} \quad \boldsymbol{R}\boldsymbol{R}^T = \boldsymbol{I} \quad \text{or} \quad \sum_{k=1}^{3} R_{ik}R_{jk} = \delta_{ij} \tag{1.88}$$

where $\boldsymbol{I}$ is an identity matrix ($I_{ij} = \delta_{ij}$)

$$\boldsymbol{I} = \begin{bmatrix} 1 & 0 & 0 \\ 0 & 1 & 0 \\ 0 & 0 & 1 \end{bmatrix} \tag{1.89}$$

This type of transformation is called an *orthogonal coordinate transformation* (OCT). A matrix $\boldsymbol{R}$ satisfying Eq. (1.88) is said to be an orthogonal matrix. That is, an orthogonal matrix is the one whose inverse is equal to the transpose. $\boldsymbol{R}$ is sometimes called a *rotation matrix*.

For example, for the coordinate rotation shown in Fig. 1.8, in 3D, the rotation matrix can be written as

$$\boldsymbol{R}=\left[\cos\,\theta\sin\theta 0-\;\sin\theta\cos\theta 0001\right] \tag{1.90}$$

In 2D, it can be written as

$$\boldsymbol{R}=\left[\cos\,\theta\sin\theta-\;\sin\theta\cos\theta\right] \tag{1.91}$$

and

$$\begin{cases} v_x=\overline{v}_x\cos\theta-\overline{v}_y\sin\theta \\ v_y=\overline{v}_x\sin\theta+\overline{v}_y\cos\theta \end{cases} \tag{1.92}$$

We recall that the determinant of a matrix product is equal to the product of the determinants. Also, the determinant of the transpose of a matrix is equal to the determinant of the matrix itself. Therefore, from Eq. (1.88), we can obtain

$$\det\left(\boldsymbol{R}\boldsymbol{R}^T\right)=(\det\boldsymbol{R})\left(\det\boldsymbol{R}^T\right)=(\det\boldsymbol{R})^2=\det\boldsymbol{I}=1 \tag{1.93}$$

and we conclude that, for an orthogonal matrix $\boldsymbol{R}$,

$$\det\boldsymbol{R}=\pm 1 \tag{1.94}$$

The plus sign occurs for rotations, and the minus sign occurs for combinations of rotations and reflections that result in a net reflection. For example, the orthogonal matrix

$$\boldsymbol{R}=\begin{bmatrix} 1 & 0 & 0 \\ 0 & 1 & 0 \\ 0 & 0 & -1 \end{bmatrix} \tag{1.95}$$

indicates a reflection in the z direction (i.e., the sign of the z component is changed).

Another property of orthogonal matrices that can be deduced directly from the definition is that the rows and columns of an orthogonal matrix must be unit vectors and mutually orthogonal. That is to say, the rows and columns form an orthonormal set.

We note that the length of a vector is unchanged under an orthogonal coordinate transformation, since the square of the length is given by

$$\vec{v}_i\vec{v}_i=R_{ij}v_jR_{ik}v_k=\delta_{jk}v_jv_k=v_jv_j \tag{1.96}$$

where the summation convention was used. That is, the square of the length of a vector is the same in both coordinate systems.

To summarize, under an orthogonal coordinate transformation, vectors are transformed according to the rule

$$\overline{\boldsymbol{v}} = \boldsymbol{R}\boldsymbol{v} \quad \text{or} \quad \overline{v}_i = \sum_{j=1}^{3} R_{ij} v_j \tag{1.97}$$

where

$$R_{ij} = \overline{\boldsymbol{e}}_i \cdot \boldsymbol{e}_j \tag{1.98}$$

and

$$\boldsymbol{R}\boldsymbol{R}^T = \boldsymbol{R}^T\boldsymbol{R} = \boldsymbol{I} \tag{1.99}$$

A vector which is transformed using an orthogonal coordinate transformation according to the rule $\overline{\boldsymbol{v}} = \boldsymbol{R}\boldsymbol{v}$ is defined as a tensor of rank 1. Examples include displacement vectors, velocity vectors, and force vectors. A tensor of rank 0 is a scalar (a quantity which is unchanged by an orthogonal coordinate transformation). For example, temperature and pressure are scalars, since $\overline{T} = T$ and $\overline{p} = p$.

We now introduce tensors of rank 2 by considering a matrix $\boldsymbol{M} = (M_{ij})$ which relates two vectors $\boldsymbol{u}$ and $\boldsymbol{v}$ by

$$\boldsymbol{v} = \boldsymbol{M}\boldsymbol{u} \quad \text{or} \quad v_i = \sum_{j=1}^{3} M_{ij} u_j \tag{1.100}$$

(i.e., the result of multiplying a matrix and a vector is a vector). Also, in a rotated coordinate system,

$$\overline{\boldsymbol{v}} = \overline{\boldsymbol{M}}\,\overline{\boldsymbol{u}} \tag{1.101}$$

Since both $\boldsymbol{u}$ and $\boldsymbol{v}$ are vectors (tensors of rank 1), Eq. (1.100) implies that

$$\boldsymbol{R}^T\overline{\boldsymbol{v}} = \boldsymbol{M}\boldsymbol{R}^T\overline{\boldsymbol{u}} \quad \text{or} \quad \overline{\boldsymbol{v}} = \boldsymbol{R}\boldsymbol{M}\boldsymbol{R}^T\overline{\boldsymbol{u}} \tag{1.102}$$

Comparing Eq. (1.101) with Eq. (1.102), we conclude that

$$\overline{\boldsymbol{M}} = \boldsymbol{R}\boldsymbol{M}\boldsymbol{R}^T \tag{1.103}$$

or, in index notation,

$$\overline{M}_{ij} = \sum_{k=1}^{3} \sum_{l=1}^{3} R_{ik} R_{jl} M_{kl} \tag{1.104}$$

which is the transformation rule for a tensor of rank 2. In general, a tensor of rank n, which has n indices, transforms under an orthogonal coordinate transformation according to the rule

$$\overline{A}_{ij\cdots k} = \sum_{p=1}^{3}\sum_{q=1}^{3}\cdots\sum_{r=1}^{3} R_{ip}R_{jq}\cdots R_{kr}A_{pq\ldots r} \tag{1.105}$$

Note that the subscripts of $\overline{\boldsymbol{A}}$ appear as the first subscripts of each $\boldsymbol{R}$ matrix, and the subscripts of $\boldsymbol{A}$ appear as the second subscripts of each $\boldsymbol{R}$ matrix. The trace of a matrix is defined as the sum of the diagonal terms. In index notation,

$$\mathrm{tr}\overline{\boldsymbol{M}} = \overline{M}_{ii} = R_{ik}R_{il}M_{kl} = \delta_{kl}M_{kl} = M_{kk} = \mathrm{tr}\boldsymbol{M} \tag{1.106}$$

which is a scalar. Thus, the trace of a matrix (a tensor of rank 2) is invariant under an orthogonal coordinate transformation. The trace is one of three invariants associated with 3 × 3 matrices.

An isotropic tensor is a tensor which is independent of the coordinate system (i.e., invariant under an orthogonal coordinate transformation). For example, the Kronecker delta δ_{ij} is an isotropic tensor, since $\overline{\delta}_{ij} = \delta_{ij}$, and

$$\overline{\delta}_{ij} = R_{ik}R_{jl}\delta_{kl} = R_{ik}R_{jk} = \delta_{ij} \tag{1.107}$$

Therefore, δ_{ij} is a second rank tensor and isotropic. In matrix notation,

$$\overline{\boldsymbol{I}} = \boldsymbol{R}\boldsymbol{I}\boldsymbol{R}^T = \boldsymbol{R}\boldsymbol{R}^T = \boldsymbol{I} \tag{1.108}$$

1.4.10 Symmetry and Skew-Symmetry

A matrix $\boldsymbol{S}$ is defined as symmetric if $\boldsymbol{S} = \boldsymbol{S}^T$, i.e., $S_{ij} = S_{ji}$. A matrix $\boldsymbol{A}$ is defined as skewsymmetric (or antisymmetric) if $\boldsymbol{A} = -\boldsymbol{A}^T$, i.e., $A_{ij} = -A_{ji}$. Note that a skew-symmetric matrix necessarily has the form

$$\boldsymbol{A} = \begin{bmatrix} 0 & A_{12} & A_{13} \\ -A_{12} & 0 & A_{23} \\ -A_{13} & -A_{23} & 0 \end{bmatrix} \tag{1.109}$$

Any matrix $\boldsymbol{M}$ can be written as the unique sum of symmetric and skew-symmetric matrices

$$\boldsymbol{M} = \boldsymbol{S} + \boldsymbol{A} \tag{1.110}$$

where

$$\boldsymbol{S} = \left(\boldsymbol{M} + \boldsymbol{M}^T\right)/2 = \boldsymbol{S}^T$$
$$\boldsymbol{A} = \left(\boldsymbol{M} - \boldsymbol{M}^T\right)/2 = -\boldsymbol{A}^T$$

For example, for a 3 × 3 matrix $\boldsymbol{M}$,

$$\boldsymbol{M} = \begin{bmatrix} 3 & 5 & 7 \\ 1 & 2 & 8 \\ 9 & 6 & 4 \end{bmatrix} = \begin{bmatrix} 3 & 3 & 8 \\ 3 & 2 & 7 \\ 8 & 7 & 4 \end{bmatrix} + \begin{bmatrix} 0 & 2 & -1 \\ -2 & 0 & 1 \\ 1 & -1 & 0 \end{bmatrix} = \boldsymbol{S} + \boldsymbol{A} \tag{1.111}$$

Note that, if $\boldsymbol{A}$ is skew-symmetric, $\boldsymbol{x} \cdot \boldsymbol{Ax} = 0$ for all $\boldsymbol{x}$. To prove this assertion, we write

$$\boldsymbol{x} \cdot \boldsymbol{Ax} = \boldsymbol{x}^T\boldsymbol{Ax} = \left(\boldsymbol{x}^T\boldsymbol{Ax}\right)^T = \boldsymbol{x}^T\,\boldsymbol{A}^T\boldsymbol{x} = -\boldsymbol{x}^T\boldsymbol{Ax} \tag{1.112}$$

where $\boldsymbol{x}^T\boldsymbol{Ax}$ is a scalar equal to its own transpose. Since this quantity is also equal to its own negative, it must vanish, and the assertion is proved. Thus, for a general matrix $\mathbf{M} = \mathbf{S} + \mathbf{A}$,

$$\boldsymbol{x} \cdot \boldsymbol{Mx} = \boldsymbol{x} \cdot \boldsymbol{Sx} \tag{1.113}$$

where $\mathbf{S}$ is the symmetric part of $\boldsymbol{M}$. The matrix product $\boldsymbol{x} \cdot \boldsymbol{Mx} = x_i M_{ij} x_j$ is a quadratic form. In expanded form, the quadratic form is

$$\begin{aligned} \boldsymbol{x} \cdot \boldsymbol{Mx} = {} & M_{11}x_1^2 + M_{22}x_2^2 + M_{33}x_3^2 \\ & + (M_{12} + M_{21})x_1x_2 + (M_{23} + M_{32})x_2x_3 + (M_{13} + M_{31})x_1x_3 \end{aligned} \tag{1.114}$$

Note that the quadratic form can be written in three different notations:

Vector (dyadic) notation :	$\boldsymbol{x} \cdot \boldsymbol{Mx}$
Matrix notation :	$\boldsymbol{x}^T\boldsymbol{Mx}$
Index notation :	$x_i M_{ij} x_j$ or $M_{ij} x_i x_j$

A matrix $\boldsymbol{M}$ is defined as positive definite if $\boldsymbol{x} \cdot \boldsymbol{Mx} > 0$ for all $\boldsymbol{x} \neq \mathbf{0}$. $\boldsymbol{M}$ is positive semi-definite if $\boldsymbol{x} \cdot \boldsymbol{Mx} \geq 0$ for all $\boldsymbol{x} \neq \mathbf{0}$.

1.4.11 Derivatives and Divergence

We consider the scalar function $\phi(x_1, x_2, x_3)$, according to the chain rule,

$$\frac{\partial \phi}{\partial \overline{x}_i} = \frac{\partial \phi}{\partial x_j} \frac{\partial x_j}{\partial \overline{x}_i} \tag{1.115}$$

where $x_j = R_{kj}\overline{x}_k$. Hence,

$$\frac{\partial x_j}{\partial \overline{x}_i} = R_{kj} \frac{\partial \overline{x}_k}{\partial \overline{x}_i} = R_{kj}\delta_{ki} = R_{ij} \tag{1.116}$$

Thus, from Eq. (1.115)

$$\frac{\partial \phi}{\partial \overline{x}_i} = R_{ij} \frac{\partial \phi}{\partial x_j} \tag{1.117}$$

from which we conclude that the partial derivative $\partial\phi/\partial x_i$ is a tensor of rank 1.

The vector operator del is defined in Cartesian coordinates as

$$\nabla = \boldsymbol{e}_1 \frac{\partial}{\partial x_1} + \boldsymbol{e}_2 \frac{\partial}{\partial x_2} + \boldsymbol{e}_3 \frac{\partial}{\partial x_3} = \boldsymbol{e}_i \frac{\partial}{\partial x_i} \tag{1.118}$$

The *gradient* of a scalar function $\phi(x_1, x_2, x_3)$ is defined as $\nabla\phi$, so that, in Cartesian coordinates,

$$\text{grad}\phi = \nabla \phi = \frac{\partial \phi}{\partial x_1}\boldsymbol{e}_1 + \frac{\partial \phi}{\partial x_2}\boldsymbol{e}_2 + \frac{\partial \phi}{\partial x_3}\boldsymbol{e}_3 = \boldsymbol{e}_\text{i} \frac{\partial \phi}{\partial x_i} \tag{1.119}$$

This vector has its components as the rates of change of ϕ with respect to the distance in the x_1, x_2 and x_3 directions, respectively. Note that, in a rotated Cartesian coordinate system,

$$\nabla \phi = \overline{\boldsymbol{e}}_i \frac{\partial \phi}{\partial \overline{x}_i} \tag{1.120}$$

The *directional* derivative measures the rate of change of a scalar function, saying $\phi(x_1, x_2, x_3)$, with respect to distance in any arbitrary direction $\boldsymbol{s}$ (Fig. 1.9) and is given by

$$\frac{\partial \phi}{\partial s} = \boldsymbol{e}_s \cdot \nabla \phi = |\boldsymbol{e}_s||\nabla \phi| \cos\theta = |\nabla \phi| \cos\theta \tag{1.121}$$

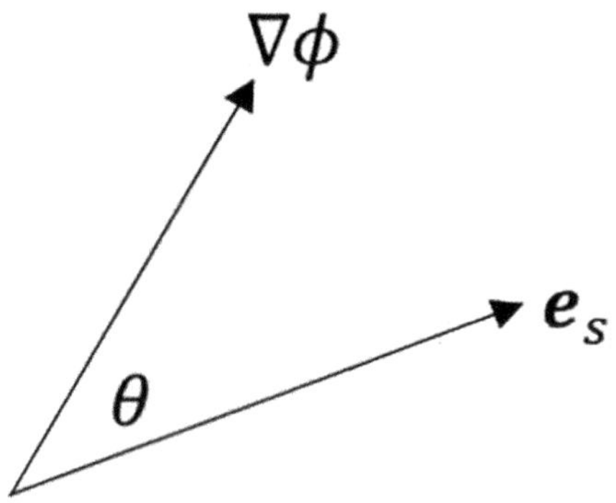

Fig. 1.9 The directional derivative

where $\boldsymbol{e}_s$ is the unit vector in the direction of $\boldsymbol{s}$, and θ is the angle between the two vectors $\nabla\phi$ and $\boldsymbol{e}_s$. Thus, the maximum change rate of ϕ is in the direction of $\nabla\phi$.

Given a vector function (field) $\boldsymbol{f}(x_1, x_2, x_3)$, the *divergence* of $\boldsymbol{f}$ is defined as $\nabla \cdot \boldsymbol{f}$, so that, in Cartesian coordinates,

$$\mathrm{div}\boldsymbol{f} = \nabla \cdot \boldsymbol{f} = \left(e_i \frac{\partial}{\partial x_i}\right) \cdot (f_j e_j) = \frac{\partial f_j}{\partial x_i}\delta_{ij} = \frac{\partial f_i}{\partial x_i} \tag{1.122}$$

or, in expanded form,

$$\nabla \cdot \boldsymbol{f} = \frac{\partial f_1}{\partial x_1} + \frac{\partial f_2}{\partial x_2} + \frac{\partial f_3}{\partial x_3} \tag{1.123}$$

We denote the partial derivative with respect to the ith Cartesian coordinate direction with the comma notation

$$\frac{\partial \phi}{\partial x_i} = \phi_{,i} \tag{1.124}$$

Using this notation, the divergence becomes (in Cartesian coordinates)

$$\nabla \cdot \boldsymbol{f} = f_{i,i} \tag{1.125}$$

The Laplacian of the scalar field $\phi(x_1, x_2, x_3)$, denoted as $\nabla^2\phi$, is defined as

$$\nabla^2\phi = \nabla \cdot \nabla\phi = \left(\boldsymbol{e}_i \frac{\partial}{\partial x_i}\right) \cdot \left(\boldsymbol{e}_j \frac{\partial \phi}{\partial x_j}\right) = \frac{\partial}{\partial x_i}\left(\frac{\partial \phi}{\partial x_j}\right)\delta_{ij} = \phi_{,ii} \tag{1.126}$$

or, in expanded form,

$$\nabla^2\phi = \frac{\partial^2\phi}{\partial x_1^2} + \frac{\partial^2\phi}{\partial x_2^2} + \frac{\partial^2\phi}{\partial x_3^2} \tag{1.127}$$

1.4.12 Even and Odd Functions

A function $f(x)$ is even if $f(-x) = f(x)$ and odd if $f(-x) = -f(x)$, as illustrated in Fig. 1.10. Geometrically, an even function is symmetrical with respect to the line $x = 0$, and an odd function is antisymmetrical. For example, $\cos x$ is even, and $\sin x$ is odd. Even and odd functions have the following useful properties:

(a) If f is even and smooth (continuous first derivative), $f'(0) = 0$.
(b) If f is odd, $f(0) = 0$.
(c) The product of two even functions is even.
(d) The product of two odd functions is even.
(e) The product of an even and an odd function is odd.
(f) The derivative of an even function is odd.
(g) The derivative of an odd function is even.
(h) If f is even, $\int_{-a}^{a} f(x)dx = 2\int_0^a f(x)dx$
(i) If f is odd, $\int_{-a}^{a} f(x)dx = 0$.

Most functions are neither even nor odd. However, any function $f(x)$ can be represented as the unique sum of even and odd functions

$$f(x) = f_e(x) + f_o(x) \tag{1.128}$$

where $f_e(x)$ is the even part of f, given by

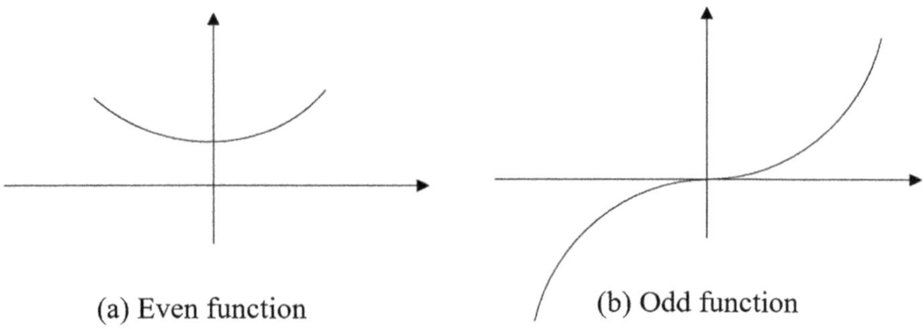

Fig. 1.10 Examples of (**a**) even and (**b**) odd functions

$$f_e(x) = \frac{1}{2}[f(x) + f(-x)] \tag{1.129}$$

and $f_o(x)$ is the odd part of f, given by

$$f_o(x) = \frac{1}{2}[f(x) - f(-x) \tag{1.130}$$

1.4.13 Index Notation and Summation Convention

In order to develop a general theory of elasticity we need to work with vectors because we live in (at least) three spatial dimensions. We can keep the notation compact by using index notation, in which we represent a vector by its components in a Cartesian (orthonormal) coordinate system. The vector $\boldsymbol{r} = (r_1, r_2, r_3)$ is represented by writing r_i where i, the index, takes the value 1,2 or 3. We also need to work with tensor quantities that are represented by matrices, with two indices:

$$\boldsymbol{A} = \begin{pmatrix} A_{11} & A_{12} \\ A_{21} & A_{22} \end{pmatrix} \tag{1.131}$$

which can be represented by $A_{\alpha\beta}$ where α and β take the values 1 or 2. We shall adopt the convention that Greek indices $(\alpha, \beta, \gamma, \ldots)$ always range from 1 to 2; and Latin indices $(i, j, k, \ldots)$ range from 1 to 3.

One advantage of index notation is that it is completely explicit. The dot product between two vectors $\boldsymbol{a}$ and $\boldsymbol{b}$ is considered, we can write this as $\boldsymbol{a} \cdot \boldsymbol{b}$ and we know (or should know) that the result is a scalar, but how do we compute it? Using index notation, we can write

$$\boldsymbol{a} \cdot \boldsymbol{b} = a_1 b_1 + a_2 b_2 + a_3 b_3 = \sum_{i=1}^{3} a_i b_i \equiv a_i b_i \tag{1.132}$$

where the last term uses the summation convention that a repeated index is summed over all values of the index, i.e. we "drop the $\sum$". Similarly, a matrix vector product $\boldsymbol{Ax} = \boldsymbol{b}$ in index notation is

$$\sum_{\beta=1}^{2} A_{\alpha\beta} x_\beta = b_\alpha \text{ or using the summation convention } A_{\alpha\beta} x_\beta = b_\alpha$$

The expression above represents two different equations because there is a so-called free (not summed) index α.

Notes: The repeated index is a dummy index because it simply represents a summation, so it can be changed when convenient

$$\sum_{i=1}^{3} a_i b_i \equiv a_i b_i = a_j b_j = a_k b_k \equiv \sum_{k=1}^{3} a_k b_k \tag{1.133}$$

- An index can never appear more than twice in a single term

$$a_i b_i c_i \text{ is NOT ALLOWED.}$$

The problem here is that if an index is repeated three times, then the meaning is potentially ambiguous. It could mean

$$\sum_{i=1}^{3} a_i b_i c_i \text{ or } \left(\sum_{i=1}^{3} (a_i b_i)\right) c_i$$

or something else. Thus, we do not allow such a construction.

- Any free (non-repeated) indices must balance (i.e. appear in every term in an equation):

$$a_i = b_i + c_i, A_{ij} = C_{ik} D_{kj}$$

The following construction is not allowed

$$a_i = A_{ij}$$

because it states that a vector is equal to a rank-two matrix, which cannot be the case. The equation $A_{ij} = B_{jk}$ is also not allowed because the free indices are not the same on both sides of the equation.

In certain circumstances we may want to repeat an index but not sum over it. In those cases, we will use brackets $e_{(i)(i)}$ means the terms e_{11}, e_{22} or e_{33} and not their sum.

The Kronecker Delta is defined by

$$\delta_{ij} = \begin{cases} 1 & \text{for} \quad i = j \\ 0 & \text{for} \quad i \neq j \end{cases}$$

It is "essentially" the identity matrix, but it can be used to exchange indices

$$a_i b_j \delta_{ij} = \sum_{i=1}^{3} \sum_{j=1}^{3} a_i b_j \delta_{ij} = a_1 b_1 \delta_{11} + a_1 b_2 \delta_{12} + a_1 b_3 \delta_{13} + a_2 b_1 \delta_{21} + a_2 b_2 \delta_{22}$$
$$+ a_2 b_3 \delta_{23} + a_3 b_1 \delta_{31} + a_3 b_2 \delta_{32} + a_3 b_3 \delta_{33} = a_1 b_1 + a_2 b_2 + a_3 b_3 = a_i b_i = a_j b_j$$

In other words, if there is a sum over one of the indices of the Kronecker Delta, the other occurrence of the index can be replaced by the remaining index in the Kronecker Delta: $a_i \delta_{ij} = a_j$. Note that $\delta_{ij}\delta_{ij} = \delta_{ii} = 3$.

To further save on writing, a comma is used to denote partial differentiation

$$\frac{\partial a}{\partial x_i} = a_{,i}, \frac{\partial u_i}{\partial x_j} = u_{i,j}$$

This means that we can express common differential operators in index notation. In Cartesian coordinates, we have:

$$\nabla \cdot \boldsymbol{u} = \text{div}\boldsymbol{u} = u_{i,i}$$
$$\nabla \phi = \text{grad}\phi = \phi_{,i}$$
$$\nabla^2 \phi = \phi_{,ii}$$

Chapter 2
Strain

Yongtao Lyu and Yonggang Zheng

2.1 Introduction

In this chapter, we begin with the development of the basic field equations of elasticity theory by first investigating the kinematics of material deformation. As a result of applied loadings, elastic solids will change shape or deform, and these deformations can be quantified by knowing the displacements of material points in the body. The continuum hypothesis establishes a displacement field at all points within the elastic solid. Using appropriate geometry, particular measures of deformation can be constructed leading to the development of the strain tensor. As expected, the strain components are related to the displacement field. The purpose of this chapter is to introduce the basic definitions of displacement and strain, to establish relations between these two field quantities, and finally to investigate the requirements to ensure single-valued and continuous displacement fields. As appropriate for linear elasticity, these kinematical results are developed under the conditions of small deformation theory. Developments in this chapter lead to two fundamental sets of field equations: the strain-displacement relations and the compatibility equations. Further field equation development, including the internal force and stress distribution, equilibrium, and elastic constitutive behavior, will be introduced in the subsequent chapters.

Strain quantifies the deformation of a solid body. It is important to realize that a solid body can move without deformation: simple translations and rotations are known as rigid-body motions and do not induce any deformation. Therefore, any measure of strain should be zero for such motions. Throughout this textbook, we only consider infinitesimal deformations.

Y. Lyu (✉) · Y. Zheng
School of Mechanics and Aerospace Engineering, Dalian University of Technology, Dalian, Liaoning, China
e-mail: Yongtaolu@dlut.edu.cn; zhengyg@dlut.edu.cn

Y. Lyu, Y. Zheng, *Theory of Elasticity*,
https://doi.org/10.1007/978-981-95-6043-1_2

2.2 Strain and displacement

Although the general relations for small deformation theory were developed in the previous section, we now want to establish a more geometrical interpretation of these results. Typically, elasticity variables and equations are field quantities defined at each point in the material continuum. However, particular field equations are often developed by first investigating the behavior of infinitesimal elements (with coordinate boundaries), and then a limiting process is invoked that allows the element to shrink to a point. The common deformational behaviors of a rectangular element are shown in Fig. 2.1. The usual types of motion include rigid-body rotation and extensional and shearing deformations as illustrated. Rigid-body motion does not contribute to the strain field, and thus also does not affect the stresses. We therefore focus our study primarily on the extensional and shearing deformation.

The strain defined in this textbook is also called the *infinitesimal strain*. It is a continuum description of the deformation. Center of mass translation and rigid rotation are not strains. The relationship between the strain and displacement can be derived as follows:

A deformable segment PAB is deformed and moved to $P'A'B'$ under an external loading (Fig. 2.2). The lengths of PA and PB are dx and dy, respectively. The coordinate of P is (x, y, z) and it moves to (x', y', z') after the deformation. The displacements in the x and y directions are represented as functions (u and v, respectively) of the coordinates. Therefore, in the deformation, the displacements of P in the x and y directions are:

$$\begin{cases} u = u(x, y, z) \\ v = v(x, y, z) \end{cases} \tag{2.1}$$

Therefore, the relationship between (x, y) and (x', y') can be written as:

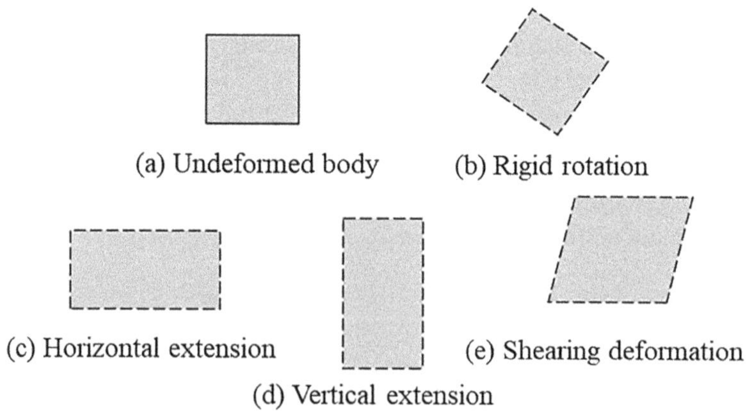

Fig. 2.1 Typical deformation of a rectangular element

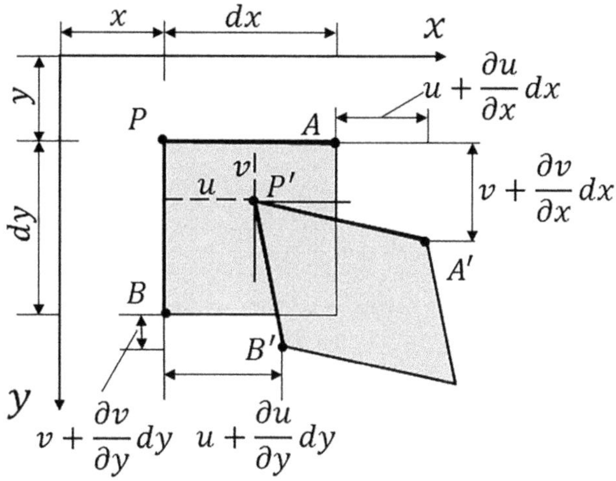

Fig. 2.2 Deformation of a plane segment

$$\begin{cases} x' = x+u(x,y,z) \\ y' = y+v(x,y,z) \end{cases} \tag{2.2}$$

The displacements at points A and B can be written as:

$$\begin{cases} u_A = u(x+dx,y,z), v_A = v(x+dx,y,z) \\ u_B = u(x,y+dy,z), v_B = v(x,y+dy,z) \end{cases} \tag{2.3}$$

Taking use of the Taylor series expansion, i.e.,

$$f(x) = \frac{f(x_0)}{0!} + \frac{f'(x_0)}{1!}(x-x_0) + \frac{f''(x_0)}{2!}(x-x_0)^2 + \ldots$$

and ignoring the higher order parts, the displacements at points A and B can be re-written as:

$$\begin{cases} u_A = u + \frac{du}{dx}dx \\ u_B = u + \frac{du}{dy}dy \end{cases} \begin{cases} v_A = v + \frac{dv}{dx}dx \\ v_B = v + \frac{dv}{dy}dy \end{cases} \tag{2.4}$$

Afterwards, the normal strains (ε_x and ε_y), defined as the relative elongations along the x and y directions (Fig. 2.2), can be written as:

$$\begin{cases} \varepsilon_x = \frac{P'A' - PA}{PA} = \frac{u + \frac{\partial u}{\partial x}dx - u}{dx} = \frac{\frac{\partial u}{\partial x}dx}{dx} = \frac{\partial u}{\partial x} \\ \varepsilon_y = \frac{P'B' - PB}{PB} = \frac{v + \frac{\partial v}{\partial y}dy - v}{dy} = \frac{\frac{\partial v}{\partial y}dy}{dy} = \frac{\partial v}{\partial y} \end{cases} \tag{2.5}$$

It can be seen from the definition that the normal strains reflect the change of the deformable body in length. The positive normal strains mean the length gets longer and the negative values imply that the length gets shorter.

The shear strain (γ_{xy}), defined as the angle change between *PA* and *PB*, can be worked out as below:

$$\tan\alpha = \frac{\left(v + \frac{\partial v}{\partial x}dx\right) - v}{dx + u + \frac{\partial u}{\partial x}dx - u} = \frac{\frac{\partial v}{\partial x}dx}{dx + \frac{\partial u}{\partial x}dx} = \frac{\frac{\partial v}{\partial x}}{1 + \varepsilon_x} = \frac{\partial v}{\partial x} \approx \alpha$$

$$\tan\beta = \frac{\left(u + \frac{\partial u}{\partial y}dy\right) - u}{dy} = \frac{\partial u}{\partial y} \approx \beta$$

then

$$\gamma_{xy} = \alpha + \beta = \frac{\partial u}{\partial y} + \frac{\partial v}{\partial x} \tag{2.6}$$

It can be seen the shear strain reflects the change of the deformable body in angle and is the consequence of the shear stress.

Combining Eq. (2.5) with Eq. (2.6), and then extend to 3D, the geometric (Cauchy's) equation in the 3D space can be written as below:

$$\begin{cases} \varepsilon_x = \dfrac{\partial u}{\partial x}, \gamma_{yz} = \dfrac{\partial w}{\partial y} + \dfrac{\partial v}{\partial z} \\ \varepsilon_y = \dfrac{\partial v}{\partial y}, \gamma_{xz} = \dfrac{\partial u}{\partial z} + \dfrac{\partial w}{\partial x} \\ \varepsilon_z = \dfrac{\partial w}{\partial z}, \gamma_{xy} = \dfrac{\partial u}{\partial y} + \dfrac{\partial v}{\partial x} \end{cases} \tag{2.7}$$

where,

$$\varepsilon_{yz} = \frac{1}{2}\gamma_{yz}, \varepsilon_{xz} = \frac{1}{2}\gamma_{xz}, \varepsilon_{xy} = \frac{1}{2}\gamma_{xy}$$

In terms of the index format

$$\varepsilon_{ij} = \frac{1}{2}\left(u_{i,j} + u_{j,i}\right) \tag{2.8}$$

Regarding the direction of the strain components, if the length of the deformable body is elongated, the normal strains are positive, i.e., ε_x, ε_y, $\varepsilon_z > 0$. If the length is shortened, the normal strains are negative, i.e., ε_x, ε_y, $\varepsilon_z < 0$. Now let's use the strain component $\gamma_{xy} = \alpha + \beta = \frac{\partial u}{\partial y} + \frac{\partial v}{\partial x}$ in Fig. 2.2 to illustrate the definition of the direction of the shear strain. First, regarding the strain component $\alpha_{yx} = \frac{\partial v}{\partial x}$, it is

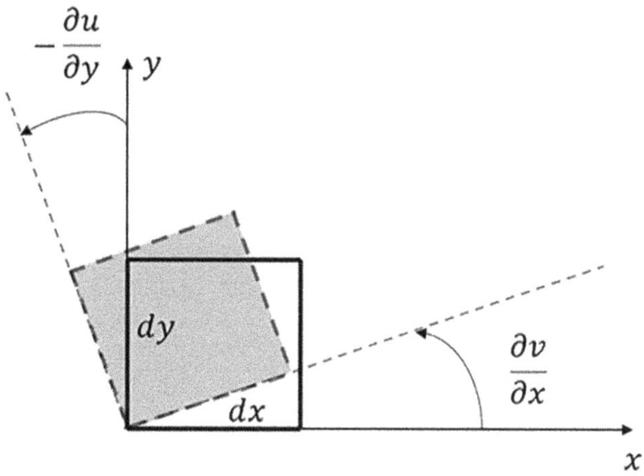

Fig. 2.3 Two-dimensional rigid-body rotation

bigger than zero, implying v increases as the increase of the coordinate x or $P'A'$ rotates towards the positive direction of y-axis. Regarding the strain component $\beta_{yx} = \frac{\partial u}{\partial y}$, it is bigger than zero, implying u increases as the increase of the coordinate y or $P'B'$ rotates towards the positive direction of x-axis.

The strain is a symmetric second-order tensor ($\varepsilon_{ij} = \varepsilon_{ji}$) and is commonly written in the matrix format:

$$\boldsymbol{\varepsilon} = [\varepsilon] = \begin{bmatrix} \varepsilon_x & \varepsilon_{xy} & \varepsilon_{xz} \\ \varepsilon_{xy} & \varepsilon_y & \varepsilon_{yz} \\ \varepsilon_{xz} & \varepsilon_{yz} & \varepsilon_z \end{bmatrix} \tag{2.9}$$

Before we conclude this geometric presentation, let's consider the rigid-body rotation of the 2D element in the $x - y$ plane, as shown in Fig. 2.3. If the element is rotated through a small rigid-body angular displacement about the z-axis, using the bottom element edge, the rotation angle is determined as $\partial v/\partial x$, while using the left edge, the angle is given by $-\partial u/\partial y$. These two expressions are of course the same; that is, $\partial v/\partial x = -\partial u/\partial y$ and note that this would imply $\varepsilon_{xy} = 0$. The rotation can then be expressed as $\omega_z = [(\partial v/\partial x) - (\partial u/\partial y)]/2$. The other components of rotation follow an analogous manner. Relations for the constant rotation ω_z can be integrated to give the result:

$$\begin{cases} u^* = u_0 - \omega_2 y \\ v^* = v_0 + \omega_2 x \end{cases} \tag{2.10}$$

where u_0 and v_0 are arbitrary constant translations in the x and y directions. This result then specifies the general form of the displacement field for 2D rigid-body motion. We can easily verify that the displacement field given by Eq. (2.10) yields zero strain. For the 3D case, the most general form of rigid-body displacement can be expressed as

$$\begin{cases} u^* = u_0 - \omega_z y + \omega_y z \\ v^* = v_0 - \omega_x z + \omega_x x \\ w^* = w_0 - \omega_y x + \omega_x y \end{cases} \tag{2.11}$$

Example 2.1: Examples of Strain and Strain Rotation
Question: To determine the displacement gradient, strain, and rotation tensors for the following displacement field: $u = \mathrm{A}x^2y$, $v = \mathrm{B}yz$, $w = \mathrm{C}xz^3$, where A, B and C are arbitrary constants. Also to calculate the dual rotation vector $\boldsymbol{\omega} = (1/2)(\nabla \times \boldsymbol{u})$.

As shown later, integrating the strain-displacement relations to determine the displacement field produces arbitrary constants and functions of integration, which are equivalent to rigid body motion terms in the form given in Eq. (2.10) or Eq. (2.11). Thus, it is important to recognize such terms. It should be noted that we normally drop these terms from the analysis because they do not contribute to the strain or stress fields.

Solution:

$$u_{i,j} = \begin{bmatrix} 2\mathrm{A}xy & \mathrm{A}x^2 & 0 \\ 0 & Bz & By \\ Cz^3 & 0 & 3Cxz^2 \end{bmatrix}$$

$$\varepsilon_{ij} = \frac{1}{2}\left(u_{i,j} + u_{j,i}\right) = \begin{bmatrix} 2Axy & \mathrm{A}x^2/2 & \mathrm{C}z^3/2 \\ Ax^2/2 & Bz & By/2 \\ Cz^3/2 & \mathrm{B}y/2 & 3Cxz^2 \end{bmatrix}$$

$$\omega_{ij} = \frac{1}{2}\left(u_{i,j} - u_{j,i}\right) = \begin{bmatrix} 0 & \mathrm{A}x^2/2 & -Cz^3/2 \\ -Ax^2/2 & 0 & By/2 \\ Cz^3/2 & -By/2 & 0 \end{bmatrix}$$

$$\boldsymbol{\omega} = \frac{1}{2}(\nabla \times \boldsymbol{u}) = \frac{1}{2}\begin{bmatrix} \boldsymbol{e}_1 & \boldsymbol{e}_2 & \boldsymbol{e}_3 \\ \partial/\partial x & \partial/\partial y & \partial/\partial z \\ Ax^2y & Byz & Cxz^3 \end{bmatrix} = \frac{1}{2}\left(-By\boldsymbol{e}_1 - Cz^3\boldsymbol{e}_2 - Ax^2\boldsymbol{e}_3\right)$$

2.3 Geometrically Nonlinear Problems

2.3.1 Classification of Geometrically Nonlinear Problems

The geometric equation presented in Eq. (2.7) is based on the assumption of linear small deformation. However, in the engineering problems, geometrically nonlinear problems are always involved, which can be classified into the following two groups:

(a) Large displacement, large rotation and small strain
In this scenario, the displacement and rotation of the deformable body are large, but the elongation and angle change are small, i.e., small strain. An example of this scenario is the deformation of the frame structure with a large span.
(b) Large displacement, large rotation and large strain
In this scenario, the displacement and rotation of the deformable body are large, and the elongation and angle change are also large. An example of this scenario is the deformation of the metal during the metal forming process.

2.3.2 Definition of Strains in Geometrically Nonlinear Problems

In the geometrically nonlinear problems, the strains and the strain-displacement relations should be derived using different theories. Figure 2.4 is used to illustrate the deformation of the deformable body in the geometrically nonlinear problem.

To describe the geometrically nonlinear problem, initial and current configurations need to be defined. The initial configuration is the configuration defined at the

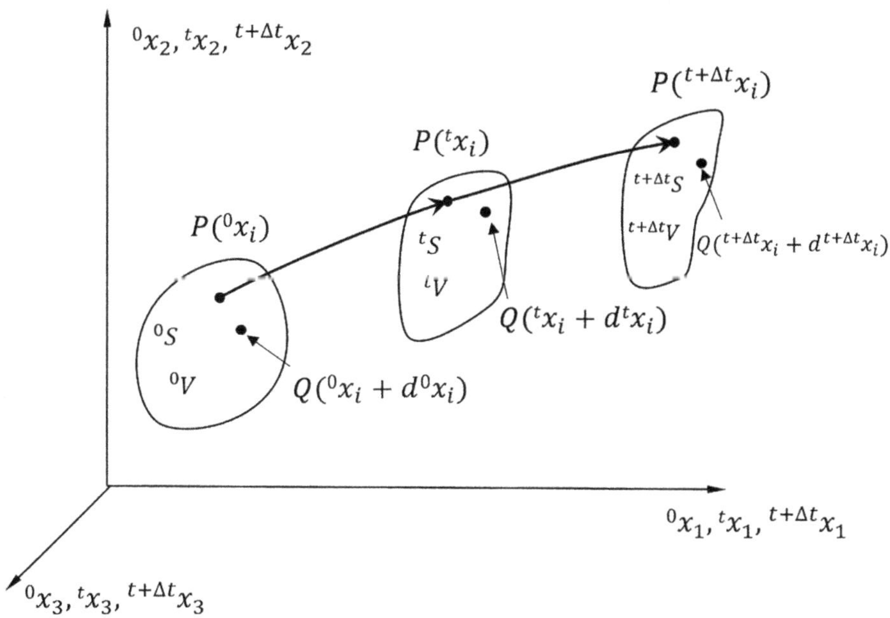

Fig. 2.4 Illustration of the configurations in geometrically nonlinear problem

initial time ($t = 0$). The coordinate of a point, point P, in the initial configuration can be written as:

$$^{0}x_i \ (i = 1, 2, 3)$$

where the number 0 at the upper left corner implies that the variable is defined at the initial configuration. The coordinate of another point, point Q, in the initial configuration can be written as

$$^{0}x_i + d^{0}x_i (i = 1, 2, 3)$$

In contrast, the current configuration is the configuration defined at time t. The coordinate of a point, i.e., point P', in the current configuration can be written as:

$$^{t}x_i \ (i = 1, 2, 3)$$

where the symbol t at the upper left corner implies that the variable is defined at the current configuration. The coordinate of another point, point Q', in the initial configuration can be written as

$$^{t}x_i + d^{t}x_i (i = 1, 2, 3)$$

The change of the locations of points from the initial configuration to the current configuration can be seen as a mathematical transformation, i.e.:

$$^{t}x_i = {}^{t}x_i\left({}^{0}x_1, {}^{0}x_2, {}^{0}x_3\right) \quad (i = 1, 2, 3)$$

Because the mapping from the initial configuration to the current configuration is one-to-one, the inverse form of the transformation can be obtained as:

$$^{0}x_i = {}^{0}x_i\left({}^{t}x_1, {}^{t}x_2, {}^{t}x_3\right) \ (i = 1, 2, 3)$$

In order to work out the strains in the geometrically nonlinear problems, the length of the line segment PQ and its change need to be worked out. To work out the relevant lengths and distances, the Lagrange and Euler coordinate systems are defined. In the Lagrange coordinate system, the coordinates at the initial configuration are taken as the unknown variables and used to describe other variables. Furthermore, the initial configuration is used to describe the deformed (current) configuration, i.e.:

$$^{t}x_i = {}^{t}x_i\left({}^{0}x_1, {}^{0}x_2, {}^{0}x_3\right)$$

In the Euler coordinate system, the coordinates at the current configuration are taken as the unknown variables to describe other variables, and the current configuration is used to describe the initial configuration, i.e.,

$$ {}^0x_i = {}^0x_i\left({}^tx_1, {}^tx_2, {}^tx_3\right) $$

Using the Lagrange coordinates, the differential of the coordinates in the current configuration can be expressed as below:

$$ d^tx_i = \left(\frac{\partial^t x_i}{\partial^0 x_j}\right) d^0x_j = {}_0^t x_{i,j} d^0x_j \tag{2.12} $$

where the letter t at the upper left corner in ${}_0^t x_{i,j}$ represents the configuration index and the number 0 at the lower left corner in ${}_0^t x_{i,j}$ represents the derivative index.

Using the Euler coordinates, the differential of the coordinate in the initial configuration can be expressed as below:

$$ d^0x_i = \left(\frac{\partial^0 x_i}{\partial^t x_j}\right) d^tx_j = {}_t^0 x_{i,j} d^tx_j \tag{2.13} $$

Then the distance from P to Q in the initial and current configurations can be written as:

$$ \left({}^0ds\right)^2 = d^0x_i d^0x_i = {}_t^0 x_{i,m} {}_t^0 x_{i,n} d^tx_m d^tx_n $$
$$ \left({}^tds\right)^2 = d^tx_i d^tx_i = {}_0^t x_{i,m} {}_0^t x_{i,n} d^0x_m d^0x_n $$

Using the Lagrange coordinates, the length change of the line segment PQ can be expressed as below:

$$ \left({}^tds\right)^2 - \left({}^0ds\right)^2 = \left({}_0^t x_{k,i} {}_0^t x_{k,j} - \delta_{ij}\right) d^0x_i d^0x_j = 2 {}_0^t \varepsilon_{ij} d^0x_i d^0x_j $$

where,

$$ {}_0^t \varepsilon_{ij} = \frac{1}{2}\left({}_0^t x_{k,i} {}_0^t x_{k,j} - \delta_{ij}\right) \tag{2.14} $$

Using the Euler coordinates, the length change of the line segment PQ can be expressed as below:

$$ \left({}^tds\right)^2 - \left({}^0ds\right)^2 = \left(\delta_{ij} - {}_t^0 x_{k,i} {}_t^0 x_{k,j}\right) d^tx_i d^tx_j = 2 {}_t^t \varepsilon_{ij} d^tx_i d^tx_j $$

where,

$$
{}_t^t\varepsilon_{ij} = \frac{1}{2}\left(\delta_{ij} - {}_t^0x_{k,i}\,{}_t^0x_{k,j}\right) \tag{2.15}
$$

After the mathematical representations for the distance of the line segment *PQ* and the length change in the two coordinate systems, the strain definition in the two systems can be given. First, using the Lagrange (initial) coordinates, the strain is given as below:

$$
{}_0^t\varepsilon_{ij} = \frac{1}{2}\left({}_0^tx_{k,i}\,{}_0^tx_{k,j} - \delta_{ij}\right) \tag{2.16}
$$

which is called the *Green-Lagrange strain tensor*, or the *Green strain tensor*.

Using the Euler (current) coordinates, the strain is given as below:

$$
{}_t^t\varepsilon_{ij} = \frac{1}{2}\left(\delta_{ij} - {}_t^0x_{k,i}\,{}_t^0x_{k,j}\right) \tag{2.17}
$$

which is called the *Almansi strain tensor*. The relation between Green and Almansi strain tensors are given as below:

$$
\begin{cases} {}_t^t\varepsilon_{ij} = {}_t^0x_{k,i}\,{}_t^0x_{l,j}\,{}_0^t\varepsilon_{kl}^{t} \\ {}_0^t\varepsilon_{ij} = {}_0^tx_{k,i}\,{}_0^tx_{l,j}\,{}_t^t\varepsilon_{kl}^{0} \end{cases} \tag{2.18}
$$

2.3.3 Definition of Deformation Gradient

Now let's come to an important definition in the geometrically nonlinear problem, i.e., the deformation gradient. To introduce the deformation gradient, let's recall the expressions for the differential of the coordinates in the two configurations.

In the current configuration,

$$
d^t x_i = \left(\frac{\partial^t x_i}{\partial^0 x_j}\right) d^0 x_j = {}_0^t x_{i,j} d^0 x_j
$$

In the initial configuration,

$$
d^0 x_i = \left(\frac{\partial^0 x_i}{\partial^t x_j}\right) d^t x_j = {}_t^0 x_{i,j} d^t x_j
$$

Then, the deformation gradient is defined as below:

$$\boldsymbol{F} = {}_0^t x_{i,j} = \frac{\partial^t x_i}{\partial^0 x_j} = \begin{bmatrix} \frac{\partial^t x_1}{\partial^0 x_1} & \frac{\partial^t x_1}{\partial^0 x_2} & \frac{\partial^t x_1}{\partial^0 x_3} \\ \frac{\partial^t x_2}{\partial^0 x_1} & \frac{\partial^t x_2}{\partial^0 x_2} & \frac{\partial^t x_2}{\partial^0 x_3} \\ \frac{\partial^t x_3}{\partial^0 x_1} & \frac{\partial^t x_3}{\partial^0 x_2} & \frac{\partial^t x_3}{\partial^0 x_3} \end{bmatrix} \tag{2.19}$$

Taking use of the chain rule, i.e.:

$$\frac{\partial}{\partial^0 x_1} = \frac{\partial}{\partial^t x_1}\frac{\partial^t x_1}{\partial^0 x_1} + \frac{\partial}{\partial^t x_2}\frac{\partial^t x_2}{\partial^0 x_1} + \frac{\partial}{\partial^t x_3}\frac{\partial^t x_3}{\partial^0 x_1}$$

The following the partial differential formulation can be obtained,

$$\begin{Bmatrix} \frac{\partial}{\partial^0 x_1} \\ \frac{\partial}{\partial^0 x_2} \\ \frac{\partial}{\partial^0 x_3} \end{Bmatrix} = \begin{bmatrix} \frac{\partial^t x_1}{\partial^0 x_1} & \frac{\partial^t x_2}{\partial^0 x_1} & \frac{\partial^t x_3}{\partial^0 x_1} \\ \frac{\partial^t x_1}{\partial^0 x_2} & \frac{\partial^t x_2}{\partial^0 x_2} & \frac{\partial^t x_3}{\partial^0 x_2} \\ \frac{\partial^t x_1}{\partial^0 x_3} & \frac{\partial^t x_2}{\partial^0 x_3} & \frac{\partial^t x_3}{\partial^0 x_3} \end{bmatrix} \begin{Bmatrix} \frac{\partial}{\partial^t x_1} \\ \frac{\partial}{\partial^t x_2} \\ \frac{\partial}{\partial^t x_3} \end{Bmatrix} \tag{2.20}$$

where,

$$\mathcal{J} = \begin{bmatrix} \frac{\partial^t x_1}{\partial^0 x_1} & \frac{\partial^t x_2}{\partial^0 x_1} & \frac{\partial^t x_3}{\partial^0 x_1} \\ \frac{\partial^t x_1}{\partial^0 x_2} & \frac{\partial^t x_2}{\partial^0 x_2} & \frac{\partial^t x_3}{\partial^0 x_2} \\ \frac{\partial^t x_1}{\partial^0 x_3} & \frac{\partial^t x_2}{\partial^0 x_3} & \frac{\partial^t x_3}{\partial^0 x_3} \end{bmatrix}$$

is called the *Jacobian matrix* and its determinant $|\mathcal{J}|$ is called the *Jacobian determinant.*

We should note that, to have a transformation for which the inverse exists, $\mathcal{J}^{-1}$ must exist, i.e.,

$$\det\mathcal{J} = |\mathcal{J}| \neq 0$$

If the body is not deformed at all, i.e., ${}^t x_i = {}^0 x_i$ $(i = 1,2,3)$, we can obtain $|\mathcal{J}| = 1$. In addition, $|\mathcal{J}| > 0$ implies for a physically realizable deformation, which can be explained as below:

The Jacobian determinant can be used to judge whether a deformation is admissible or not. For example, we can represent the displacements of a point from the initial configuration to the deformed configuration as below:

$$\begin{cases} u_1 = {}^t x_1 - {}^0 x_1 \\ u_2 = {}^t x_2 - {}^0 x_2 \\ u_3 = {}^t x_3 - {}^0 x_3 \end{cases} \tag{2.21}$$

from which we can work out

$$\frac{\partial^t x_1}{\partial^0 x_1} = 1 + \frac{\partial u_1}{\partial^0 x_1}, \frac{\partial^t x_2}{\partial^0 x_1} = \frac{\partial u_2}{\partial^0 x_1}, \ldots$$

Then

$$|\mathcal{J}| = \begin{vmatrix} \frac{\partial^t x_1}{\partial^0 x_1} & \frac{\partial^t x_2}{\partial^0 x_1} & \frac{\partial^t x_3}{\partial^0 x_1} \\ \frac{\partial^t x_1}{\partial^0 x_2} & \frac{\partial^t x_2}{\partial^0 x_2} & \frac{\partial^t x_3}{\partial^0 x_2} \\ \frac{\partial^t x_1}{\partial^0 x_3} & \frac{\partial^t x_2}{\partial^0 x_3} & \frac{\partial^t x_3}{\partial^0 x_3} \end{vmatrix} = \begin{vmatrix} 1+\frac{\partial u_1}{\partial^0 x_1} & \frac{\partial u_2}{\partial^0 x_1} & \frac{\partial u_3}{\partial^0 x_1} \\ \frac{\partial u_1}{\partial^0 x_2} & 1+\frac{\partial u_2}{\partial^0 x_2} & \frac{\partial u_3}{\partial^0 x_2} \\ \frac{\partial u_1}{\partial^0 x_3} & \frac{\partial u_2}{\partial^0 x_3} & 1+\frac{\partial u_3}{\partial^0 x_3} \end{vmatrix} > 0$$

Therefore, for the admissible deformation described, $|\mathcal{J}| > 0$. In reverse, if $|\mathcal{J}| > 0$, then we can say the deformation is admissible.

In summary, from the Jacobian determinant, the deformation state of deformable body can be determined, i.e., $|\mathcal{J}| = 1$ implying the body is undeformed and $|\mathcal{J}| > 0$ implying the body is physically deformed.

For example, considering the displacement field

$$\begin{cases} u_1 = {}^t x_1 - {}^0 x_1 = {}^0 x_1 - 2{}^0 x_2 \\ u_2 = {}^t x_2 - {}^0 x_2 = 3{}^0 x_1 + 2{}^0 x_2 \\ u_3 = {}^t x_3 - {}^0 x_3 = 5{}^0 x_3 \end{cases}$$

for which

$$|\mathcal{J}| = \begin{vmatrix} 2 & 3 & 0 \\ -2 & 3 & 0 \\ 0 & 0 & 6 \end{vmatrix} = 72 > 0$$

Thus, the displacement field is admissible.

The deformation gradient and the Jacobian determinant are important concepts in the geometrically nonlinear problems due to the physical meaning associated with it. Below let's use the 2D case to explore their physical meaning. In 2D case shown in Fig. 2.5, the (x_1, x_2) and the (ξ_1, ξ_2) systems are defined. When x_1 is increased by Δx_1 which is projected to the ξ_1 and ξ_2 axes, the following relations can be obtained:

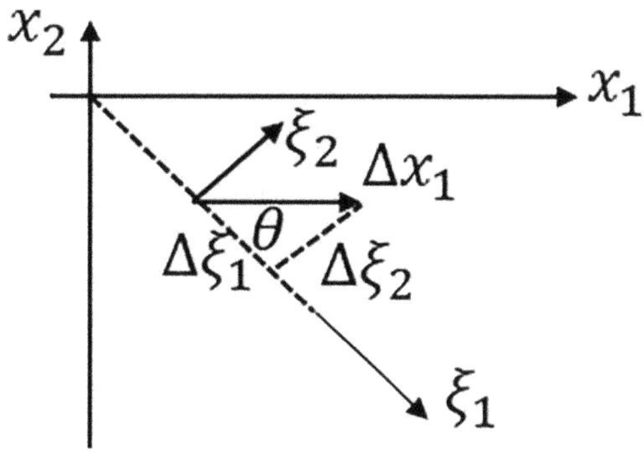

Fig. 2.5 Illustration of the physical meaning of deformation gradient in 2D case

$$\frac{\Delta\xi_1}{\Delta x_1} = \cos\theta, \frac{\Delta\xi_2}{\Delta x_1} = \sin\theta$$

It can be seen $\begin{bmatrix} \frac{\Delta\xi_1}{\Delta x_1} & \frac{\Delta\xi_2}{\Delta x_1} \end{bmatrix}$ is the vector which is tangent to the x_1 curve in the (ξ_1, ξ_2) system.

Similarly, in the 3D scenario,

$$d\boldsymbol{X}_1 = \begin{Bmatrix} \frac{\partial^t x_1}{\partial^0 x_1} \\ \frac{\partial^t x_2}{\partial^0 x_1} \\ \frac{\partial^t x_3}{\partial^0 x_1} \end{Bmatrix} dx_1, d\boldsymbol{X}_2 = \begin{Bmatrix} \frac{\partial^t x_1}{\partial^0 x_2} \\ \frac{\partial^t x_2}{\partial^0 x_2} \\ \frac{\partial^t x_3}{\partial^0 x_2} \end{Bmatrix} dx_2, d\boldsymbol{X}_3 = \begin{Bmatrix} \frac{\partial^t x_1}{\partial^0 x_3} \\ \frac{\partial^t x_2}{\partial^0 x_3} \\ \frac{\partial^t x_3}{\partial^0 x_3} \end{Bmatrix} dx_3 \quad (2.22)$$

Extending the conclusions from 2D to 3D, it can be concluded that $d\boldsymbol{X}_1$, $d\boldsymbol{X}_2$, $d\boldsymbol{X}_3$ are the three vectors which are tangent to the ${}^0x_1, {}^0x_2, {}^0x_3$ curves in the $({}^tx_1, {}^tx_2, {}^tx_3)$ coordinate system. In Eq. (2.22), $|d\boldsymbol{X}_1|$, $|d\boldsymbol{X}_2|$, $|d\boldsymbol{X}_3|$ are the magnitudes of changes in the three directions.

The volume of the 3D deformable body in the $({}^tx_1, {}^tx_2, {}^tx_3)$ system after the deformation can be written as:

$$dV = d\boldsymbol{X}_1 \cdot (d\boldsymbol{X}_2 \times d\boldsymbol{X}_3) = \begin{vmatrix} \frac{\partial^t x_1}{\partial^0 x_1} & \frac{\partial^t x_2}{\partial^0 x_1} & \frac{\partial^t x_3}{\partial^0 x_1} \\ \frac{\partial^t x_1}{\partial^0 x_2} & \frac{\partial^t x_2}{\partial^0 x_2} & \frac{\partial^t x_3}{\partial^0 x_2} \\ \frac{\partial^t x_1}{\partial^0 x_3} & \frac{\partial^t x_2}{\partial^0 x_3} & \frac{\partial^t x_3}{\partial^0 x_3} \end{vmatrix} d^0x_1 d^0x_2 d^0x_3 = |\mathcal{J}| d^0x_1 d^0x_2 d^0x_3$$

where $V = d^0x_1 d^0x_2 d^0x_3$ is the volume before the deformation.

Therefore,

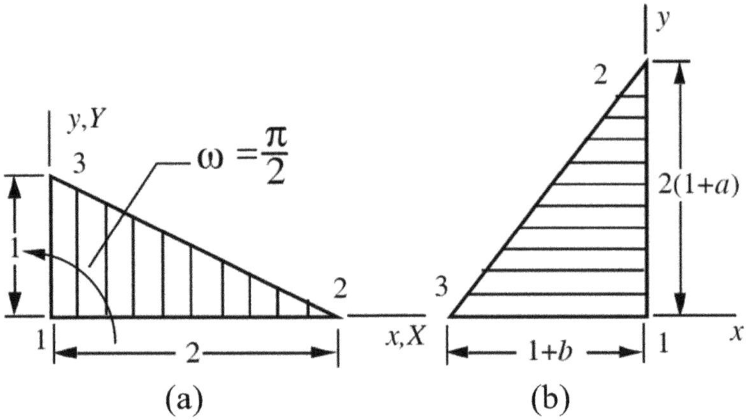

Fig. 2.6 Rotation and stretching of a triangular: (**a**) initial configuration; (**b**) deformed configuration

$$|\mathcal{J}| = \frac{V + \Delta V}{V} = 1 + \frac{\Delta V}{V}$$

where, $\frac{\Delta V}{V}$ is the volumetric strain. The equation above gives the physical meaning of the Jacobian determinant, i.e., $|\mathcal{J}|$ is the ratio of the new volume to the old volume for the admissible deformation of a 3D deformable body.

Example 2.2: Rotation and Stretching of a Triangular
Question: A triangular is rotated and stretched (Fig. 2.6). The nodal coordinates of the triangular are represented as $(x_1, \ y_1)$, $(x_2, \ y_2)$ and $(x_3, \ y_3)$. At time t, the nodal coordinates are expressed as a function of the time t as below:

$$^{t}x_1 = {^{t}y_1} = 0$$

$$^{t}x_2 = 2(1 + at)\cos\frac{\pi t}{2}$$

$$^{t}y_2 = 2(1 + at)\sin\frac{\pi t}{2}$$

$$^{t}x_3 = -(1 + bt)\sin\frac{\pi t}{2}$$

$$^{t}y_3 = (1 + bt)\cos\frac{\pi t}{2}$$

The task is to work out the deformation gradient.

Solution: At time t, the coordinate of a point within the triangular can be represented as:

$$\begin{cases} {}^t x(\xi, t) = {}^t x_1 \xi_1 + {}^t x_2 \xi_2 + {}^t x_3 \xi_3 \\ {}^t y(\xi, t) = {}^t y_1 \xi_1 + {}^t y_2 \xi_2 + {}^t y_3 \xi_3 \end{cases}$$

where (x_1, y_1), (x_2, y_2), (x_3, y_3) are the coordinates of the three nodes of the triangular, ξ_1, ξ_2 and ξ_3 are the weighting factors.

At time zero, i.e., $t = 0$, the coordinates worked out should be equal to the initial coordinates of the triangular, i.e.:

$$\begin{cases} {}^0 x = {}^t x(\xi, 0) \\ {}^0 y = {}^t y(\xi, 0) \end{cases}$$

From Fig. 2.6a, it can be noted that the x coordinates of nodes 1 and 3 are equal to each other, the y coordinates of nodes 1 and 2 are equal to each other, the x coordinate of node 2 is 2.0 and the y coordinate of node 3 is 1.0, i.e.,

$$\begin{cases} {}^0 x_1 = {}^0 x_3, {}^0 x_2 = 2 \\ {}^0 y_1 = {}^0 y_2, {}^0 y_3 = 1 \end{cases}$$

Putting these values back to the equations, we can work out:

$$\xi_2 = \frac{1}{2} {}^0 x, \xi_3 = {}^0 y$$

Because the origin of the coordinate system is at node 1 in both the initial and deformed configurations, ${}^0 x_1 = {}^0 y_1 = 0$, ${}^t x_1 = {}^t y_1 = 0$. Therefore, the coordinates of the point at time t can be rewritten as:

$$\begin{cases} {}^t x = {}^0 x(1 + at) \cos \dfrac{\pi t}{2} - {}^0 y(1 + bt) \sin \dfrac{\pi t}{2} \\ {}^t y = {}^0 x(1 + at) \sin \dfrac{\pi t}{2} + {}^0 y(1 + bt) \cos \dfrac{\pi t}{2} \end{cases}$$

From which, we can work out the deformation gradient as below:

$$\boldsymbol{F} = {}_0^t x_{i,j} = \begin{bmatrix} \dfrac{\partial^t x}{\partial^0 x} & \dfrac{\partial^t x}{\partial^0 y} \\ \dfrac{\partial^t y}{\partial^0 x} & \dfrac{\partial^t y}{\partial^0 y} \end{bmatrix} = \begin{bmatrix} (1 + at) \cos \dfrac{\pi t}{2} & -(1 + bt) \sin \dfrac{\pi t}{2} \\ (1 + at) \sin \dfrac{\pi t}{2} & (1 + bt) \cos \dfrac{\pi t}{2} \end{bmatrix}$$

In this chapter, we have introduced the definition of some new strains. In the Mechanics of Materials, the strain definitions, such as the engineering strain, are introduced. Then the differences between different strains should be noted and appropriate strain should be used in different engineering problems. An example below is used to illustrate the results of different strains in the stretch of a 1D bar.

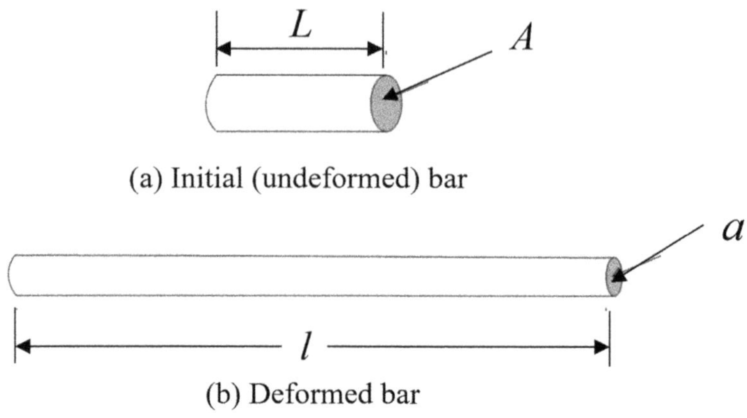

Fig. 2.7 Deformation of the 1D bar: (**a**) undeformed bar; (**b**) deformed bar

Example 2.3: Comparison of the Strains in 1D Problem
Question: A 1D bar with the initial length of L and cross-sectional area of A is shown in Fig. 2.7. The bar is stretched and at time t, its length is l and cross-sectional area is a. The task is to work out the engineering strain, the Green strain and the Almansi strain.

Solution: Using ${}^0x$ and ${}^tx$ to represent the x coordinates of the same point within the bar at the initial and current configurations, respectively, and then the displacement of this point can be worked out as:

$$ {}^tu = {}^tx - {}^0x = \frac{l}{L}{}^0x - {}^0x = {}^tx - \frac{L}{l}{}^tx $$

Then based on strain definitions, the engineering strain can be worked out as:

$$ \varepsilon_E = \frac{1}{2}\left(\frac{\partial^t u}{\partial^0 x} + \frac{\partial^t u}{\partial^0 x}\right) = \frac{l-L}{L} $$

The Green strain can be worked out as:

$$ \varepsilon_G = {}_0^t\varepsilon = \frac{1}{2}\left(\frac{\partial^t u}{\partial^0 x} + \frac{\partial^t u}{\partial^0 x} + \frac{\partial^t u}{\partial^0 x}\frac{\partial^t u}{\partial^0 x}\right) = \frac{l^2 - L^2}{2L^2} $$

The Almansi strain can be worked out as:

$$ \varepsilon_A = {}_t^t\varepsilon = \frac{1}{2}\left(\frac{\partial^t u}{\partial^t x} + \frac{\partial^t u}{\partial^t x} + \frac{\partial^t u}{\partial^t x}\frac{\partial^t u}{\partial^t x}\right) = \frac{l^2 - L^2}{2l^2} $$

To make the comparisons, the three strains are plotted in Fig. 2.8. It is shown that when the deformation is small, i.e., the l/L is around 1.0, the three strains almost

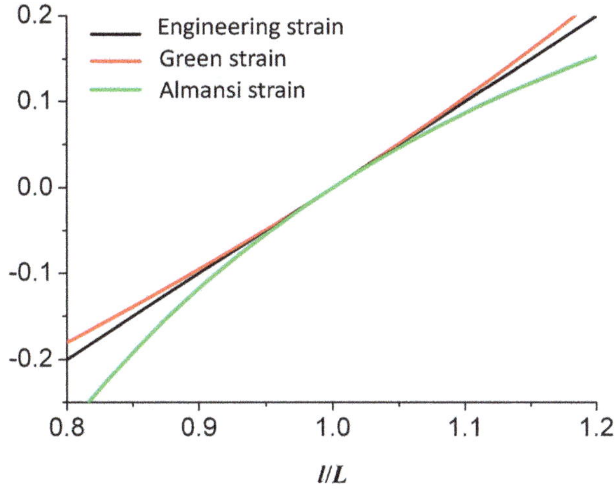

Fig. 2.8 Comparison of different strains calculated from the same problem

overlap with each other. When the deformation becomes large (in the stretching case, $l/L > 1.1$ and in the shortening case, $l/L < 0.9$), the differences among the three strains become prominent. It is also revealed in Fig. 2.8 that the larger the deformation is, the larger the differences are. These phenomenon matches with the definitions of the strains. The engineering strain is defined in the small deformation and is deemed to be used in the small deformation. In contrast, the Green and Almansi strains are defined and deemed to be used in the large deformation scenario. Therefore, in the large deformation scenario, the Engineering strain produces the errors when it is used to describe the strain. It also should be noted that because the Green strain is defined using the Lagrange coordinate and the Almansi strain is defined using the Euler coordinate, there are of course discrepancies between them in the large deformation scenario. Therefore, to accurately describe the strain in the large deformation scenario, the correct selection from the Green or Almansi strain should be made.

2.4 Relative Displacement Tensor

When external loading is applied, elastic solids would deform. Example of a simple 2D cantilever beam is shown in Fig. 2.9. The rectangular beam in the vertical position (Fig. 2.9a) is taken as the undeformed configuration, and the deformed shape/configuration is shown in Fig. 2.9b. As it is typical in most problems, the deformation varies from point to point and is thus said to be nonhomogeneous. A superimposed square mesh is shown in the two configurations, and this indicates how elements within the material will deform locally. It is apparent that elements within the mesh undergo extensional and shearing deformation. An elastic solid is said to be deformed or strained when the relative displacements between points in

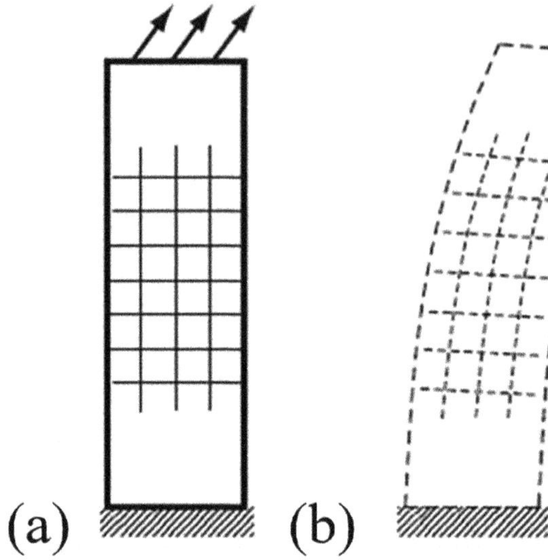

Fig. 2.9 Example of 2D deformation: (**a**) undeformed beam; (**b**) deformed beam

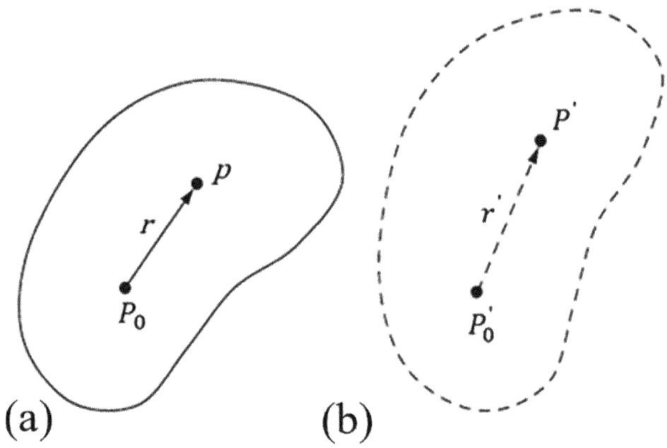

Fig. 2.10 General deformation between two neighboring points: (**a**) undeformed configuration; (**b**) deformed configuration

the body are changed. This is in contrast to rigid-body motion where the distance between points remains the same.

In order to quantify deformation, the general example shown in Fig. 2.10 is considered. In the undeformed configuration, we identify two neighboring material points P_0 and P connected with the relative position vector $\boldsymbol{r}$ as shown in Fig. 2.10. Through a general deformation, these points are mapped to the locations P_0' and P' in the deformed configuration. For finite or large deformation theory, the undeformed and deformed configurations can be significantly different, and a distinction between these two configurations must be maintained leading to Lagrangian and Eulerian descriptions. However, since we are developing linear elasticity, which uses only small deformation theory, the distinction between undeformed and deformed configurations can be dropped.

Using Cartesian coordinates, the displacement vectors of points P_0 and P are defined to be u^0 and u, respectively. Since P and P_0 are neighboring points, we can use a Taylor series expansion around point P_0 to express the components of $\boldsymbol{u}$ as

$$\begin{cases} u = u^0 + \dfrac{\partial u}{\partial x} r_x + \dfrac{\partial u}{\partial y} r_y + \dfrac{\partial u}{\partial z} r_z \\ v = v^0 + \dfrac{\partial v}{\partial x} r_x + \dfrac{\partial v}{\partial y} r_y + \dfrac{\partial v}{\partial z} r_z \\ w = w^0 + \dfrac{\partial w}{\partial x} r_x + \dfrac{\partial w}{\partial y} r_y + \dfrac{\partial w}{\partial z} r_z \end{cases} \tag{2.23}$$

where u, v, w are the Cartesian components of the displacement vector.

Note that the higher-order terms of the expansion have been dropped since the components of $\boldsymbol{r}$ are small. The change in the relative position vector $\boldsymbol{r}$ can be written as

$$\Delta \boldsymbol{r} = \boldsymbol{r}' - \boldsymbol{r} = \boldsymbol{u} - \boldsymbol{u}^0$$

and using Eq. (2.23) gives

$$\begin{cases} \Delta r_x = \dfrac{\partial u}{\partial x} r_x + \dfrac{\partial u}{\partial y} r_y + \dfrac{\partial u}{\partial z} r_z \\ \Delta r_y = \dfrac{\partial v}{\partial x} r_x + \dfrac{\partial v}{\partial y} r_y + \dfrac{\partial v}{\partial z} r_z \\ \Delta r_z = \dfrac{\partial w}{\partial x} r_x + \dfrac{\partial w}{\partial y} r_y + \dfrac{\partial w}{\partial z} r_z \end{cases}$$

or in index notation

$$\Delta r_i = u_{i,j} r_j$$

The tensor $u_{i,j}$ is called the *displacement gradient tensor*, and may be written out as

$$u_{i,j} = \begin{bmatrix} \dfrac{\partial u}{\partial x} & \dfrac{\partial u}{\partial y} & \dfrac{\partial u}{\partial z} \\ \dfrac{\partial v}{\partial x} & \dfrac{\partial v}{\partial y} & \dfrac{\partial v}{\partial z} \\ \dfrac{\partial w}{\partial x} & \dfrac{\partial w}{\partial y} & \dfrac{\partial w}{\partial z} \end{bmatrix} \tag{2.24}$$

From Eq. (2.24), this tensor can be decomposed into symmetric and antisymmetric parts as

$$u_{i,j} = \varepsilon_{ij} + \omega_{ij} \tag{2.25}$$

where,

$$\begin{cases} \varepsilon_{ij} = \frac{1}{2}\left(u_{i,j} + u_{j,i}\right) \\ \omega_{ij} = \frac{1}{2}\left(u_{i,j} - u_{j,i}\right) \end{cases}$$

The tensor e_{ij} is called the *strain tensor*, while ω_{ij} is referred to as the *rotation tensor*. Equation (2.25) thus imply that for small deformation theory, the change in the relative position vector between neighboring points can be expressed in terms of a sum of strain and rotation components. Combining Eq. (2.23) with Eq. (2.25), and choosing $r_i = dx_i$, we can also write the general result in the form

$$u_i = u_i^0 + \varepsilon_{ij}dx_j + w_{ij}dx_j$$

Because we are considering a general displacement field, these results include both strain deformation and rigid-body motion. Recall that a dual vector w_i can be associated with the rotation tensor such that $\omega_i = -\varepsilon_{ijk}\omega_{jk}/2$. Using this definition, it is found that

$$\begin{cases} \omega_1 = \omega_{32} = \frac{1}{2}\left(\frac{\partial u_3}{\partial x_2} - \frac{\partial u_2}{\partial x_3}\right) \\ \omega_2 = \omega_{13} = \frac{1}{2}\left(\frac{\partial u_1}{\partial x_3} - \frac{\partial u_3}{\partial x_1}\right) \\ \omega_3 = \omega_{21} = \frac{1}{2}\left(\frac{\partial u_2}{\partial x_1} - \frac{\partial u_1}{\partial x_2}\right) \end{cases} \tag{2.26}$$

which can be expressed collectively in vector format as $\boldsymbol{\omega} = (1/2)(\nabla \times \boldsymbol{u})$. As it is shown in the next section, these components represent rigid-body rotation of material elements about the coordinate axes. These general results indicate that the strain deformation is related to the strain tensor e_{ij}, which in turn is related to the displacement gradients. We next pursue a more geometric approach and determine specific connections between the strain tensor components and geometric deformation of material elements.

2.4.1 Physical Meaning of the Strain and Rotation Tensors

A differential line segment in Fig. 2.11 is used to further illustrate the physical meaning of the strain and rotation tensors. The line segment is deformed from the initial position AB to the finial position $A'B'$. Coordinates of points A and B are written as:

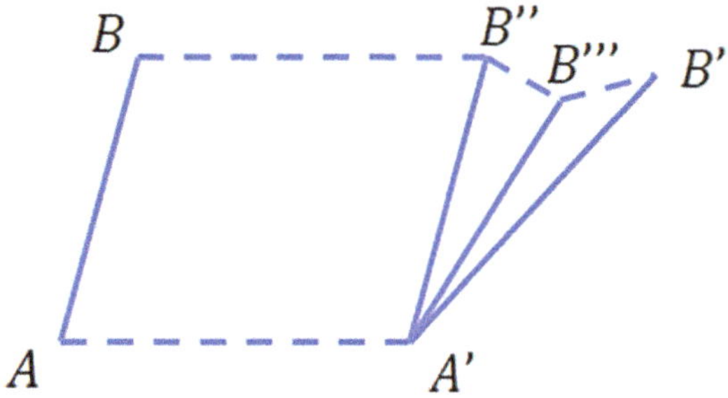

Fig. 2.11 Illustration of the physical meaning of the strain and rotation tensors

$$A:(x,y,z),B:(x+dx,y+dy,z+dz)$$

The displacement from A to $A^{'}$ is represented as:

$$u(x,y,z),v(x,y,z),w(x,y,z)$$

Then the displacements from B to $B^{'}$ can be written as:

$$\begin{cases} u^{'}=u(x+dx,y+dy,z+dz) \\ v^{'}=v(x+dx,y+dy,z+dz) \\ w^{'}=w(x+dx,y+dy,z+dz) \end{cases}$$

Using the Taylor series expansion and ignoring the high-order terms (because of small deformation), the displacements from B to $B^{'}$ can be rewritten as:

$$\begin{cases} u^{'}=u+\frac{\partial u}{\partial x}dx+\frac{\partial u}{\partial y}dy+\frac{\partial u}{\partial z}dz \\ v^{'}=v+\frac{\partial v}{\partial x}dx+\frac{\partial v}{\partial y}dy+\frac{\partial v}{\partial z}dz \\ w^{'}=w+\frac{\partial w}{\partial x}dx+\frac{\partial w}{\partial y}dy+\frac{\partial w}{\partial z}dz \end{cases} \tag{2.27}$$

The relations above can be rewritten as:

$$\begin{cases} u^{'}=u+\varepsilon_x dx+\frac{1}{2}\gamma_{xy}dy+\frac{1}{2}\gamma_{xz}dz-\frac{1}{2}\omega_z dy+\frac{1}{2}\omega_y dz \\ v^{'}=v+\frac{1}{2}\gamma_{xy}\ dx+\varepsilon_y dy+\frac{1}{2}\gamma_{yz}dz+\frac{1}{2}\omega_z dx-\frac{1}{2}\omega_x dz \\ w^{'}=w+\frac{1}{2}\gamma_{xz}dx+\frac{1}{2}\gamma_{yz}dy+\varepsilon_z dz-\frac{1}{2}\omega_y dx+\frac{1}{2}\omega_x dy \end{cases}$$

In terms of matrix format, the relations above can be written as:

$$\begin{bmatrix} u' \\ v' \\ w' \end{bmatrix} = \begin{bmatrix} u \\ v \\ w \end{bmatrix} + \begin{bmatrix} 0 & -\frac{1}{2}\omega_z & \frac{1}{2}\omega_y \\ \frac{1}{2}\omega_z & 0 & -\frac{1}{2}\omega_x \\ -\frac{1}{2}\omega_y & \frac{1}{2}\omega_x & 0 \end{bmatrix} \begin{bmatrix} dx \\ dy \\ dz \end{bmatrix}$$

$$+ \begin{bmatrix} \varepsilon_x & \frac{1}{2}\gamma_{xy} & \frac{1}{2}\gamma_{xz} \\ \frac{1}{2}\gamma_{xy} & \varepsilon_y & \frac{1}{2}\gamma_{yz} \\ \frac{1}{2}\gamma_{xz} & \frac{1}{2}\gamma_{yz} & \varepsilon_z \end{bmatrix} \begin{bmatrix} dx \\ dy \\ dz \end{bmatrix} \quad (2.28)$$

where, the first term in the left side of the equation, i.e. $[u\ v\ w]^T$ is due to the rigid translation from AB to $A'B''$. The second terms, i.e.,

$$\begin{bmatrix} 0 & -\frac{1}{2}\omega_z & \frac{1}{2}\omega_y \\ \frac{1}{2}\omega_z & 0 & -\frac{1}{2}\omega_x \\ -\frac{1}{2}\omega_y & \frac{1}{2}\omega_x & 0 \end{bmatrix} \begin{bmatrix} dx \\ dy \\ dz \end{bmatrix} = \omega_{ij} \begin{bmatrix} dx \\ dy \\ dz \end{bmatrix} \quad (2.29)$$

is to describe the rigid rotation from $A'B''$ to $A'B'''$, where ω_{ij} is the rigid rotation tensor. The third terms, i.e.,

$$\begin{bmatrix} \varepsilon_x & \frac{1}{2}\gamma_{xy} & \frac{1}{2}\gamma_{xz} \\ \frac{1}{2}\gamma_{xy} & \varepsilon_y & \frac{1}{2}\gamma_{yz} \\ \frac{1}{2}\gamma_{xz} & \frac{1}{2}\gamma_{yz} & \varepsilon_z \end{bmatrix} \begin{bmatrix} dx \\ dy \\ dz \end{bmatrix} = \varepsilon_{ij} \begin{bmatrix} dx \\ dy \\ dz \end{bmatrix} \quad (2.30)$$

is to describe the pure deformation (strain) from $A'B'''$ to $A'B'$, where ε_{ij} is the infinitesimal or Cauchy strain tensor.

In summary, Eq. (2.28) has decomposed the deformation from AB to $A'B'$ into three parts, the rigid translation from AB to $A'B''$, the rigid rotation from $A'B''$ to $A'B'''$, and the pure deformation from $A'B'''$ to $A'B'$.

2.5 Strain Transformation

Because the strains are components of a second-order tensor, the transformation theory can be applied. Applying the transformation relation to the strain gives

$$\varepsilon'_{ij} = Q_{ip}Q_{jq}\varepsilon_{pq} \quad (2.31)$$

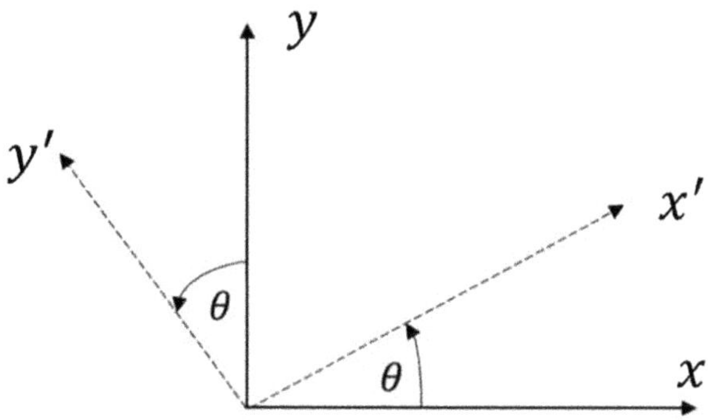

Fig. 2.12 Two-dimensional rotational transformation

where the rotation matrix $Q_{ij} = \cos(x_i', x_j)$. Thus, given the strain in one coordinate system, we can determine the new components in any other rotated system. For the general 3D case, define the rotation matrix as

$$Q_{ij} = \begin{bmatrix} l_1 & m_1 & n_1 \\ l_2 & m_2 & n_2 \\ l_3 & m_3 & n_3 \end{bmatrix}$$

Using this notational scheme, the specific transformation relations of Eq. (2.31) become

$$\left\{ \begin{array}{c} \varepsilon_x' = \varepsilon_x l_1^2 + \varepsilon_y m_1^2 + \varepsilon_z n_1^2 + 2(\varepsilon_{xy} l_1 m_1 + \varepsilon_{yz} m_1 n_1 + \varepsilon_{zx} n_1 l_1) \\ \varepsilon_y' = \varepsilon_x l_2^2 + \varepsilon_y m_2^2 + \varepsilon_z n_2^2 + 2(\varepsilon_{xy} l_2 m_2 + \varepsilon_{yz} m_2 n_2 + \varepsilon_{zx} n_2 l_2) \\ \varepsilon_z' = \varepsilon_x l_3^2 + \varepsilon_y m_3^2 + \varepsilon_z n_3^2 + 2(\varepsilon_{xy} l_3 m_3 + \varepsilon_{yz} m_3 n_3 + \varepsilon_{zx} n_3 l_3) \\ \varepsilon_{xy}' = \varepsilon_x l_1 l_2 + \varepsilon_y m_1 m_2 + \varepsilon_z n_1 n_2 + \varepsilon_{xy}(l_1 m_2 + m_1 l_2) + \varepsilon_{yz}(m_1 n_2 + n_1 m_2) \\ + \varepsilon_{zx}(n_1 l_2 + l_1 n_2) \\ \varepsilon_{yz}' = \varepsilon_x l_2 l_3 + \varepsilon_y m_2 m_3 + \varepsilon_z n_2 n_3 + \varepsilon_{xy}(l_2 m_3 + m_2 l_3) + \varepsilon_{yz}(m_2 n_3 + n_2 m_3) \\ + \varepsilon_{zx}(n_2 l_3 + l_2 n_3) \\ \varepsilon_{zx}' = \varepsilon_x l_3 l_1 + \varepsilon_y m_3 m_1 + \varepsilon_z n_3 n_1 + \varepsilon_{xy}(l_3 m_1 + m_3 l_1) + \varepsilon_{yz}(m_3 n_1 + n_3 m_1) \\ + \varepsilon_{zx}(n_3 l_1 + l_3 n_1) \end{array} \right. \tag{2.32}$$

For the 2D case shown in Fig. 2.12, the transformation matrix can be expressed as

$$Q_{ij} = [\cos\theta \sin\theta 0 - \sin\theta \cos\theta 0 0 0 1]$$

Using this transformation, the in-plane strain components can be transformed according to

$$\left\{ \begin{array}{l} \varepsilon_x' = \varepsilon_x \cos^2\theta + \varepsilon_y \sin^2\theta + 2\varepsilon_{xy} \sin\theta \cos\theta \\ \varepsilon_y' = \varepsilon_x \sin^2\theta + \varepsilon_y \cos^2\theta - 2\varepsilon_{xy} \sin\theta \cos\theta \\ \varepsilon_{xy}' = -\varepsilon_x \sin\theta \cos\theta + \varepsilon_y \sin\theta \cos\theta + \varepsilon_{xy}(\cos^2\theta - \sin^2\theta) \end{array} \right. \tag{2.33}$$

which is commonly rewritten in the double angle format:

$$\begin{cases} \varepsilon_x^{'} = \dfrac{\varepsilon_x + \varepsilon_y}{2} + \dfrac{\varepsilon_x - \varepsilon_y}{2} \cos 2\theta + \varepsilon_{xy} \sin 2\theta \\ \varepsilon_y^{'} = \dfrac{\varepsilon_x + \varepsilon_y}{2} - \dfrac{\varepsilon_x - \varepsilon_y}{2} \cos 2\theta - \varepsilon_{xy} \sin 2\theta \\ \varepsilon_{xy}^{'} = \dfrac{\varepsilon_y - \varepsilon_x}{2} \sin 2\theta + \varepsilon_{xy} \cos 2\theta \end{cases}$$

Transformation relations defined in Eq. (2.33) can be directly applied to establish the transformations between Cartesian and polar coordinate systems. Additional applications of these results can be found when dealing with experimental strain gauge measurement systems. For example, standard experimental methods using a rosette strain gauge allow the determination of extensional strains in three different directions on the surface of a structure. Using this type of data, Eq. (2.33) can be repeatedly used to establish three independent equations that can be solved for the state of strain (ε_x, ε_y, ε_{xy}) at the surface point under study.

2.6 Calculation of Principal Strains

Because the strain is a symmetric second-order tensor, we can identify and determine its principal axes and values. According to this theory, for any given strain tensor we can establish the principal value problem and solve the characteristic equation to explicitly determine the principal values and directions. The general characteristic equation for the strain tensor can be written as

$$\det\left[\varepsilon_{ij} - \varepsilon\delta_{ij}\right] = -\varepsilon^3 + I_1\varepsilon^2 - I_2\varepsilon + I_3 = 0 \tag{2.34}$$

where ε is the principal strain and the fundamental invariants of the strain tensor can be expressed in terms of the three principal strains ε_1, ε_2, ε_3 as

$$\begin{cases} I_1 = \varepsilon_1 + \varepsilon_2 + \varepsilon_3 \\ I_2 = \varepsilon_1\varepsilon_2 + \varepsilon_2\varepsilon_3 + \varepsilon_3\varepsilon_1 \\ I_3 = \varepsilon_1\varepsilon_2\varepsilon_3 \end{cases} \tag{2.35}$$

The first invariant $I_1 = I$ is normally called the *cubical dilatation*, because it is related to the change in volume of material elements. The strain matrix in the principal coordinate system takes the special diagonal form

$$\varepsilon_{ij} = \begin{bmatrix} \varepsilon_1 & 0 & 0 \\ 0 & \varepsilon_2 & 0 \\ 0 & 0 & \varepsilon_3 \end{bmatrix}$$

Notice that in this principal coordinate system, the deformation does not produce any shearing and thus is only extensional. Therefore, a rectangular element oriented along principal axes of strain will retain its orthogonal shape and undergo only extensional deformation of its sides.

2.7 Some Strain Definitions

2.7.1 Spherical and Deviatoric Strains

In particular applications it is convenient to decompose the strain tensor into two parts called *spherical and deviatoric strain tensors*. The spherical strain is defined as

$$\widetilde{\varepsilon}_{ij} = \frac{1}{3}\varepsilon_{kk}\delta_{ij} = \frac{1}{3}\vartheta\delta_{ij} \tag{2.36}$$

while the deviatoric strain is specified as

$$\hat{\varepsilon}_{ij} = \varepsilon_{ij} - \frac{1}{3}\varepsilon_{kk}\delta_{ij} \tag{2.37}$$

Note that the total strain is then simply the sum of spherical and deviatoric strains, i.e.,

$$\varepsilon_{ij} = \widetilde{\varepsilon}_{ij} + \hat{\varepsilon}_{ij} \tag{2.38}$$

The spherical strain represents only volumetric deformation and is an isotropic tensor, being the same in all coordinate systems. The deviatoric strain tensor then accounts for changes in the shape of material elements. It can be shown that the principal directions of the deviatoric strain are the same as those of the strain tensor.

2.7.2 Volumetric Strains

A differential hexahedron with the dimension of dx, dy and dz (Fig. 2.13) is considered. Before the deformation, the volume of the hexahedron can be written as:

$$V = dxdydz$$

Fig. 2.13 Illustration of volumetric strain

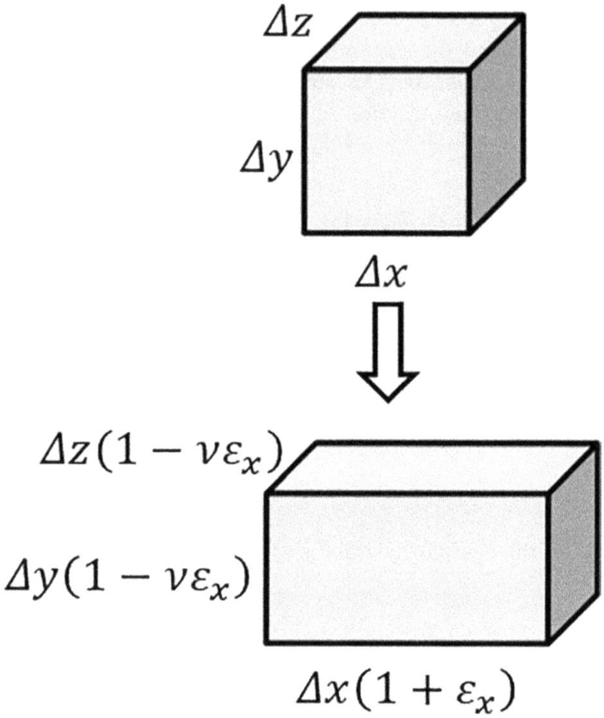

and the volume after deformation is:

$$V^* = dx(1+\varepsilon_x)dy(1+\varepsilon_y)dz(1+\varepsilon_z)$$
$$\approx dxdydz(1+\varepsilon_x+\varepsilon_y+\varepsilon_z)$$

The volumetric strain is defined as the change in the volume divided by the initial volume:

$$\theta = \frac{V^* - V}{V} = (\varepsilon_x + \varepsilon_y + \varepsilon_z) \tag{2.39}$$

or

$$\theta = \frac{\partial u}{\partial x} + \frac{\partial v}{\partial y} + \frac{\partial w}{\partial z}$$

It can be seen the volumetric strain is equal to the first invariant I_1. Second, when the volumetric strain is larger than zero, i.e., $\theta > 0$, it implies that the volume of the deformable body is expanding. On the contrary, when the volumetric strain is smaller than zero, i.e., $\theta < 0$, it implies the volume of the deformable body is shrinking.

2.8 Strain Compatibility

We now investigate in more details the nature of the strain-displacement relations, and this will lead to the development of some additional relations necessary to ensure continuous and single-valued displacement field solutions. The geometric equations (Eq. (2.7)) and the index notation in Eq. (2.8) represent six equations for the six strain components in terms of three displacements. If we specify continuous and single-valued displacements, u, v, w, then through differentiation, the resulting strain field will be equally well behaved. However, the converse is not necessarily true; that is, given the six strain components, integration of the strain-displacement relations (Eq. (2.7)) does not necessarily produce continuous and single-valued displacements. This should not be totally surprising since we are trying to solve six equations for only three unknown displacement components. In order to ensure continuous and single-valued displacements, the strains must satisfy additional relations called *integrability* or *compatibility equations.*

Before we proceed with the mathematics to develop these equations, it is instructive to consider a geometric interpretation of this concept. A 2D example is shown in Fig. 2.14, whereby an elastic solid is first divided into a series of elements in case (a). For simple visualization, only four such elements are considered. In the undeformed configuration shown in case (b), these elements of course fit together perfectly. Next, let's arbitrarily specify the strain of each of the four elements and attempt to reconstruct the solid. For case (c), the elements have been carefully strained, taking neighboring elements into consideration so that the system fits together thus yielding continuous and single-valued displacements. However, for case (d), the elements have been individually deformed without any concern for neighboring deformations. It is observed for this case that the system will not fit together without voids and gaps, and this situation produces a discontinuous displacement field. Therefore, it is concluded that the strain components must be somehow related to yield continuous and single-valued displacements. We now pursue these particular relations.

The process of developing these equations is based on eliminating the displacements from the strain-displacement relations. Working in index notation, we start by differentiating Eq. (2.8) twice with respect to x_k and x_l:

$$\varepsilon_{ij,kl} = \frac{1}{2}\left(u_{i,jkl} + u_{j,ikl}\right) \tag{2.40}$$

Through simple interchange of subscripts, we can generate the following additional relations:

$$\begin{cases} \varepsilon_{kl,ij} = \dfrac{1}{2}\left(u_{k,lij} + u_{l,kij}\right) \\ \varepsilon_{jl,ik} = \dfrac{1}{2}\left(u_{j,lik} + u_{l,jik}\right) \\ \varepsilon_{ik,jl} = \dfrac{1}{2}\left(u_{i,kjl} + u_{k,ijl}\right) \end{cases}$$

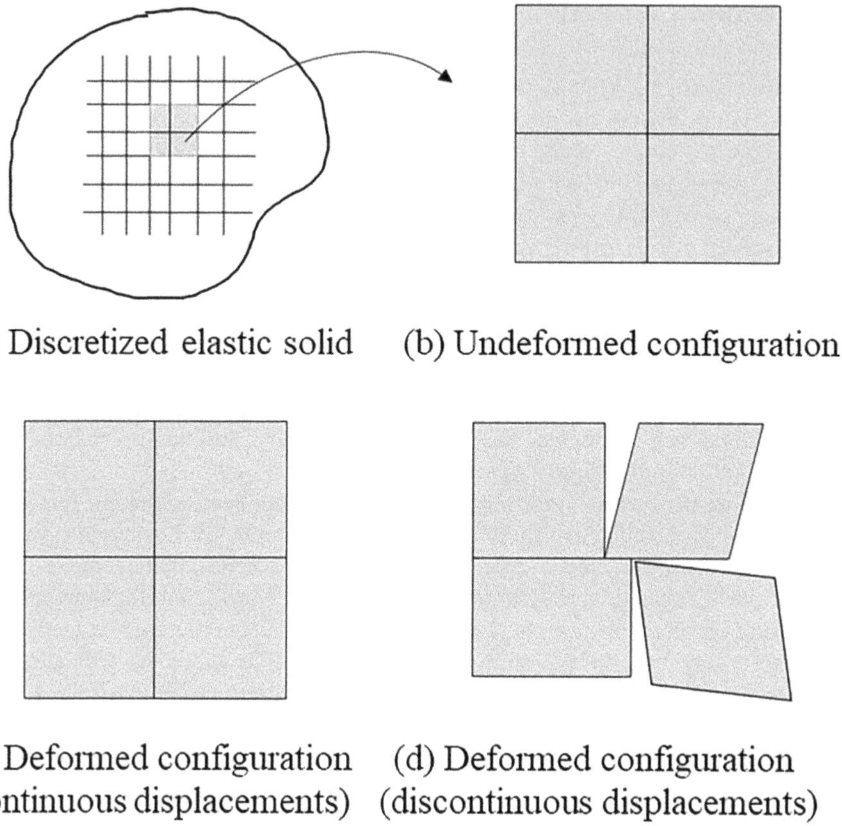

Fig. 2.14 Physical interpretation of strain compatibility

Working under the assumption of continuous displacements, we can interchange the order of differentiation on $\boldsymbol{u}$ and the displacements can be eliminated from the preceding set to get

$$\varepsilon_{ij,kl} + \varepsilon_{kl,ij} - \varepsilon_{ik,jl} - \varepsilon_{jl,ik} = 0 \tag{2.41}$$

These are called the *Saint-Venant compatibility equations*. Although the system would lead to 81 individual equations, most are either simple identities or repetitions, and only 6 are meaningful. These six relations may be determined by letting $k = l$, and in scalar notation, they become

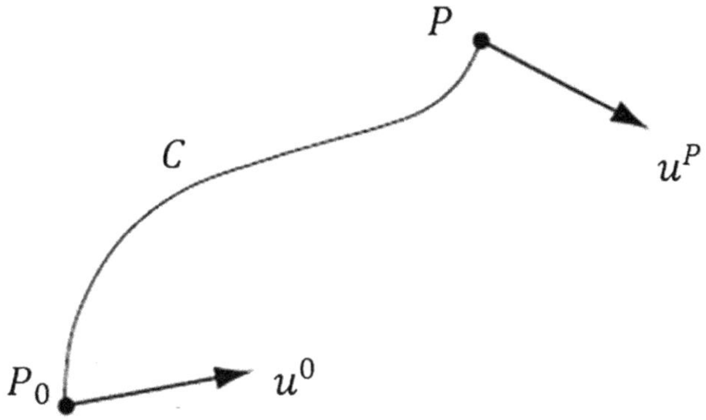

Fig. 2.15 Continuity of displacement

$$\left\{\begin{aligned} \frac{\partial^2\varepsilon_x}{\partial y^2}+\frac{\partial^2\varepsilon_y}{\partial x^2}&=2\frac{\partial^2\varepsilon_{xy}}{\partial x\partial y}\\ \frac{\partial^2\varepsilon_y}{\partial z^2}+\frac{\partial^2\varepsilon_z}{\partial y^2}&=2\frac{\partial^2\varepsilon_{yz}}{\partial y\partial z}\\ \frac{\partial^2\varepsilon_z}{\partial x^2}+\frac{\partial^2\varepsilon_x}{\partial z^2}&=2\frac{\partial^2\varepsilon_{zx}}{\partial z\partial x}\\ \frac{\partial^2\varepsilon_x}{\partial y\partial z}&=\frac{\partial}{\partial x}\left(-\frac{\partial\varepsilon_{yz}}{\partial x}+\frac{\partial\varepsilon_{zx}}{\partial y}+\frac{\partial\varepsilon_{xy}}{\partial z}\right)\\ \frac{\partial^2\varepsilon_y}{\partial z\partial x}&=\frac{\partial}{\partial y}\left(-\frac{\partial\varepsilon_{zx}}{\partial y}+\frac{\partial\varepsilon_{xy}}{\partial z}+\frac{\partial\varepsilon_{yz}}{\partial x}\right)\\ \frac{\partial^2\varepsilon_z}{\partial x\partial y}&=\frac{\partial}{\partial z}\left(-\frac{\partial\varepsilon_{xy}}{\partial z}+\frac{\partial\varepsilon_{yz}}{\partial x}+\frac{\partial\varepsilon_{zx}}{\partial y}\right) \end{aligned}\right. \tag{2.42}$$

It can be seen that these six equations are not all independent. In the development of the compatibility relations, we assume that the displacements are continuous, and thus the resulting equations (Eq. (2.42)) are actually only a necessary condition. In order to show that they are also sufficient, two arbitrary points P and P_0 in an elastic solid, as shown in Fig. 2.15, are considered, without loss in generality, the origin may be placed at point P_0.

The displacements of points P and P_0 are denoted by u_i^P and u_i^0, and the displacement of point P can be expressed as

$$u_i^P=u_i^0+\int_c du_i=u_i^0+\int_c \frac{\partial u_i}{\partial x_j}dx_j \tag{2.43}$$

where C is any continuous curve connecting points P_0 and P. Using Eq. (2.25) for the displacement gradient, Eq. (2.43) becomes

$$u_i^P = u_i^0 + \int_c (e_{ij} + \omega_{ij}) dx_j \tag{2.44}$$

Integrating the last term by parts gives

$$\int_C \omega_{ij} dx_j = \omega_{ij}^P x_j^P - \int_C x_j \omega_{ij,k} dx_k \tag{2.45}$$

where ω_{ij}^P is the rotation tensor at point P. Using expression for the rotation tensor in Eq. (2.25), we can have

$$w_{ij,k} = \frac{1}{2}\left(u_{i,jk} - u_{j,ik}\right) = \frac{1}{2}\left(u_{i,jk} - u_{j,ik}\right) + \frac{1}{2}\left(u_{k,ji} - u_{k,ji}\right) = \frac{1}{2}\frac{\partial}{\partial x_j}$$
$$\times \left(u_{i,k} + u_{k,i}\right) - \frac{1}{2}\frac{\partial}{\partial x_i}\left(u_{j,k} + u_{k,j}\right) = \varepsilon_{ik,j} - \varepsilon_{jk,i} \tag{2.46}$$

Substituting Eqs. (2.45) and (2.46) into Eq. (2.44) yields

$$u_i^P = u_i^P + \omega_{ij}^P x_j^P + \int_c U_{ik} dx_k \tag{2.47}$$

where $U_{ik} = \varepsilon_{ik} - x_j(\varepsilon_{ik,\ j} - \varepsilon_{jk,\ i})$.

Now if the displacements are to be continuous and single-valued functions, the line integral appearing in Eq. (2.47) must be the same for any curve C; that is, the integral must be independent of the path of integration. This implies that the integrand must be an exact differential, so that the value of the integral depends only on the endpoints. Invoking Stokes theorem, we can show that if the region is simply connected (definition of the term simply connected is postponed for the moment), a necessary and sufficient condition for the integral to be path independent is for $U_{ik,\ l} = U_{il,\ k}$. Using this result yields

$$\varepsilon_{ik,l} - \delta_{jl}\left(\varepsilon_{ik,j} - \varepsilon_{jk,i}\right) - x_j\left(\varepsilon_{ik,jl} - \varepsilon_{jk,il}\right) = \varepsilon_{il,k} - \delta_{jk}\left(\varepsilon_{il,j} - \varepsilon_{j,i}\right) - x_j\left(\varepsilon_{il,jk} - \varepsilon_{jl,ik}\right)$$

which reduces to

$$x_j\left(\varepsilon_{ik,jl} - \varepsilon_{jk,il} - \varepsilon_{il,jk} + \varepsilon_{jl,ik}\right) = 0$$

Because this equation must be true for all values of x_j, the terms in parentheses must vanish, and after some index renaming, this gives the identical result previously stated by the compatibility relations (Eq. (2.41)):

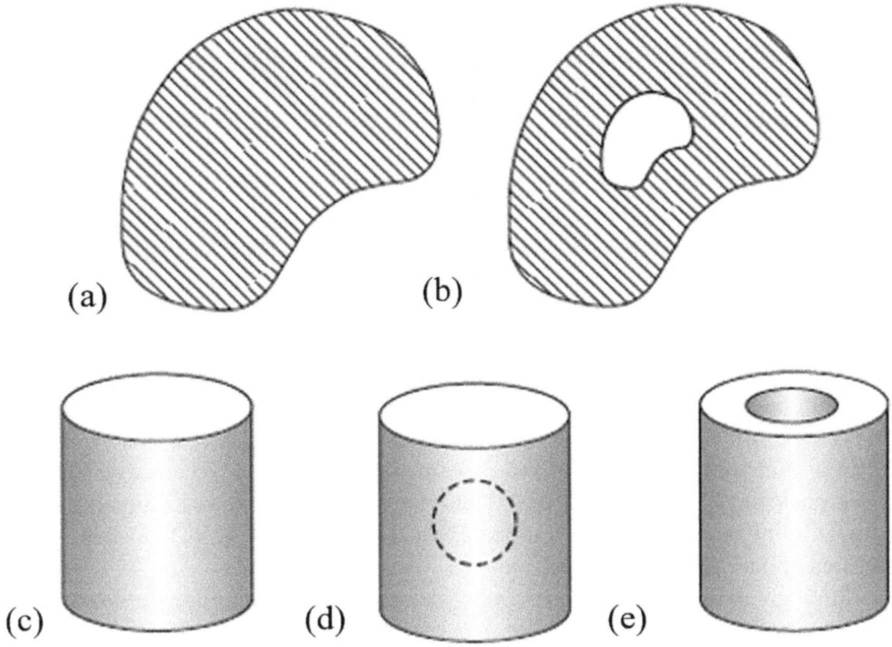

Fig. 2.16 Examples of domain connectivity: (**a**) 2D simply connected domain; (**b**) 2D multiply connected domain; (**c**) 3D simply connected domain; (**d**) 3D simply connected domain; (**e**) 3D multiply connected domain

$$\varepsilon_{ij,kl} + \varepsilon_{kl,ij} - \varepsilon_{ik,jl} - \varepsilon_{jl,ik} = 0$$

Thus, the compatibility relations (Eq. (2.41) or Eq. (2.42)) are the necessary and sufficient conditions for continuous and single-valued displacements in simply connected regions. Now let's get back to the term simply connected. This concept is related to the topology or geometry of the region under study. There are several places in elasticity theory where the connectivity of the region fundamentally affects the formulation and solution method. The term simply connected refers to regions of space for which all simple closed curves drawn in the region can be continuously shrunk to a point without going outside the region. Domains not having this property are called multiply connected. Several examples of such regions are illustrated in Fig. 2.16. A general simply connected 2D region is shown in Fig. 2.16a, and clearly this case allows any contour within the region to be shrunk to a point without going out of the domain. However, if we create a hole in the region as shown in Fig. 2.16b, a closed contour surrounding the hole cannot be shrunk to a point without going into the hole and thus outside of the region. Thus, for 2D regions, the presence of one or more holes makes the region multiply connected. Note that by introducing a cut between the outer and inner boundaries in Fig. 2.16b, a new region is created that is now simply connected. Thus, multiply connected regions can be made simply connected by introducing one or more cuts between appropriate boundaries. Figure 2.16c illustrates a simply connected 3D example of a solid circular cylinder. If a spherical cavity is placed inside this cylinder as shown in Fig. 2.16d, the region is still simply connected because any closed contour can still be shrunk to a point by sliding around the interior cavity. However, if the cylinder has a through hole as shown in Fig. 2.16e, then an interior contour encircling the axial through hole cannot be reduced to a point without going into the hole and outside the body. Thus, Fig. 2.16e is an example of the multiply connected 3D region.

The compatibility equations (Eq. (2.41) or (2.42) are necessary and sufficient conditions for continuous and single-valued displacements only for simply connected regions. However, for multiply connected domains, compatibility relations provide not only necessary but also sufficient conditions. For this case, further relations can be developed and imposed on the problem, and these are found through the introduction of cuts within the region to make it simply connected as per our earlier discussion. For 2D case shown in Fig. 2.16b, this process will lead to a relation commonly called a *Cesaro integral* taken around any closed irreducible curve enclosing the internal cavity. Thus, for multiply connected domains, strain compatibility is guaranteed if compatibility relations (Eq. (2.41) or (2.42)) are satisfied and all *Cesarro integrals* vanish.

Although the compatibility relations guarantee (under appropriate conditions) continuous displacements, they do not ensure uniqueness of the displacement field. It should be noted that, through integration of the strain displacement relations, the displacements can be determined only up to an arbitrary rigid-body motion. In some elasticity problems (e.g. thermal stress, crack problems and dislocation modeling), it is necessary to use multi-valued displacement fields to properly model the problem.

Chapter 3
Stress

Yongtao Lyu and Yonggang Zheng

3.1 Introduction

In the previous chapter, the kinematics of deformation without regard to the force or stress distribution within the elastic solid is investigated and discussed. We now want to examine these issues and explore the transmission of forces through deformable materials. Our study leads to the definition and use of the traction vector and stress tensor. Each provides a quantitative method to describe both boundary and internal force distributions within a continuum solid. Because it is commonly accepted that maximum stresses are a major contributing factor to material failure, primary application of elasticity theory is used to determine the distribution of stress within a given structure. Related to these force distribution issues is the concept of equilibrium. Within a deformable solid, the force distribution at each point must be balanced. For the static case, the summation of forces on an infinitesimal element is required to be zero, while for a dynamic problem the resultant force must be equal to the mass times the element's acceleration. In this chapter, we establish the definitions and properties of the traction vector and stress tensor and develop the equilibrium equations, which become another set of field equations necessary in the overall formulation of elasticity theory. It should be noted that the developments in this chapter do not require the material to be elastic, and thus in principle these results can be applied to a broader class of material behavior.

Y. Lyu (✉) · Y. Zheng
School of Mechanics and Aerospace Engineering, Dalian University of Technology, Dalian, Liaoning, China
e-mail: Yongtaolu@dlut.edu.cn; zhengyg@dlut.edu.cn

Y. Lyu, Y. Zheng, *Theory of Elasticity*,
https://doi.org/10.1007/978-981-95-6043-1_3

3.2 Some Definitions Related to Stress

3.2.1 Body and Surface Forces

When a structure is subjected to applied external loadings, internal forces are induced inside the body. Following the theory from continuum mechanics, these internal forces are distributed continuously within the solid. In order to investigate such forces, it is convenient to categorize them into two major groups, commonly referred to as body forces and surface forces.

Body forces are proportional to the body's mass and are reacted with an agent outside of the body. Examples of these include gravitational-weight forces, magnetic forces, and inertial forces. Fig. 3.1a shows an example of the body force of an object's self-weight. By using the principles from continuum mechanics, a body force density $\boldsymbol{F}(x)$ is the force per unit volume that acts on the mass inside a body. The resultant body force on a material volume V is the integral of the body force density over the body volume:

$$\boldsymbol{F}_R = \iiint_V \boldsymbol{F}(x) dV \tag{3.1}$$

Surface forces always act on a surface and result from physical contact with another body. Figure 3.1b illustrates surface forces existing in a beam section that has been created by sectioning the body into two pieces. For this particular case, the surface S is a virtual one in the sense that it is artificially created to investigate the nature of the internal forces at this location in the body. Again, the resultant surface force over the entire surface S can be expressed as the integral of a surface force density function $\boldsymbol{T}^n(x)$ over the acting surface:

$$\boldsymbol{F}_S = \iint_S \boldsymbol{T}^n(x) dS \tag{3.2}$$

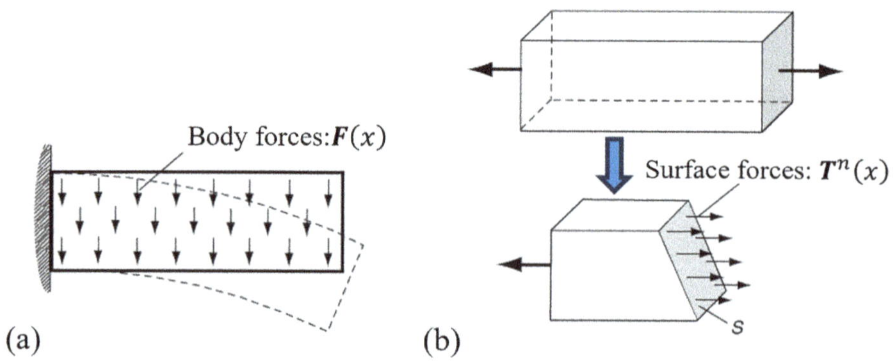

Fig. 3.1 Examples of body and surface forces: (**a**) Cantilever beam under self-weight; (**b**) Sectioned axially loaded beam

The surface force density is normally referred to as the traction vector and is discussed in more details in the next section. In the development of classical elasticity, distributions of body or surface couples are normally not included. Theories that consider such force distributions have been constructed in an effort to extend classical elasticity for applications in micromechanical modeling.

3.2.2 Traction Vector and Stress Tensor

In order to quantify the nature of the internal distribution of forces within a continuum solid, a general body subject to arbitrary (concentrated and distributed) external loadings, as shown in Fig. 3.2, is considered. To investigate the internal forces, a section is made through the body (Fig. 3.2). On this section, considering a small area ΔA with a unit normal vector $\boldsymbol{n}$, the resultant surface force acting on ΔA is defined by $\Delta \boldsymbol{F}$. Consistent with our earlier discussion, no resultant surface couple is included. The stress or traction vector is defined as

$$\boldsymbol{T}^n(x, \boldsymbol{n}) = \lim_{\Delta A \to 0} \frac{\Delta \boldsymbol{F}}{\Delta A} \tag{3.3}$$

Notice that the traction vector depends on both the spatial location and the unit normal vector to the surface under study. Thus, even though we may be investigating the same point, the traction vector still varies as a function of the orientation of the surface normal. Because the traction is defined as force per unit area, the total surface force is determined through integration as per Eq. (3.2). Note, also, the simple action-reaction principle (Newton's third law) is given as below:

$$\boldsymbol{T}^n(x, \boldsymbol{n}) = -\boldsymbol{T}^n(x, -\boldsymbol{n}) \tag{3.4}$$

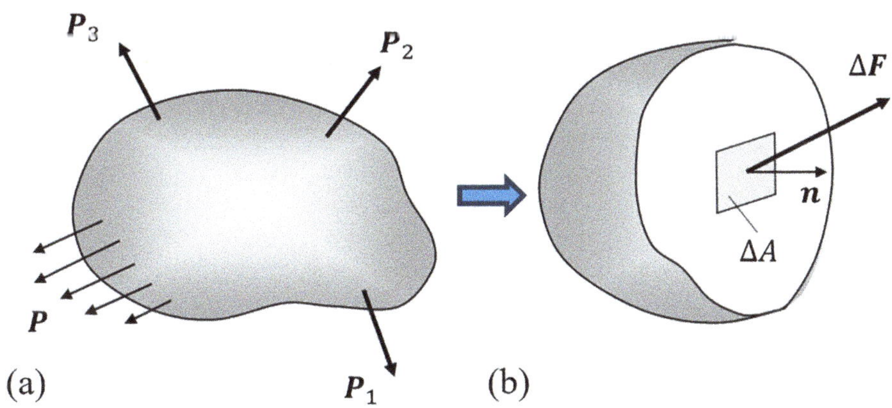

Fig. 3.2 Sectioned solid under external loading: (**a**) externally loaded body; (**b**) sectioned body

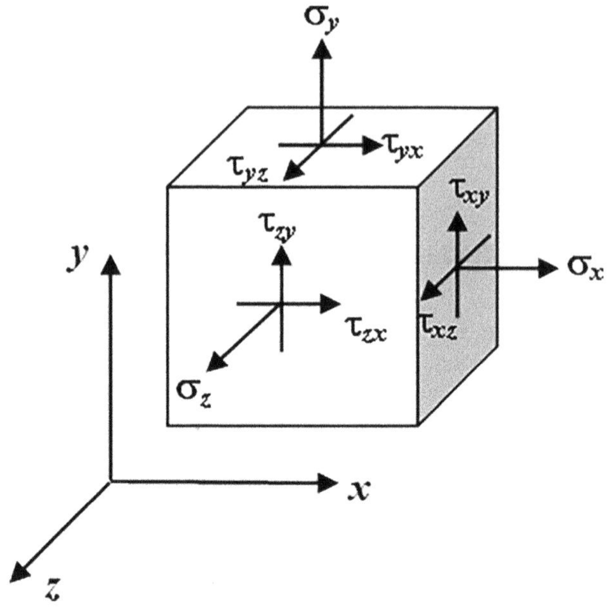

Fig. 3.3 Stress components in the three orthogonal planes

We now consider the special case in which ΔA coincides with each of the three coordinate planes with the unit normal vectors pointing along the positive coordinate axes. This concept is shown in Fig. 3.3, where the three coordinate surfaces for ΔA partitioning off a cube of material. For this case, the traction vector on each face can be written as

$$\begin{cases} \boldsymbol{T}^n(x, \boldsymbol{n} = \boldsymbol{e}_1) = \sigma_x \boldsymbol{e}_1 + \tau_{xy} \boldsymbol{e}_2 + \tau_{xz} \boldsymbol{e}_3 \\ \boldsymbol{T}^n(x, \boldsymbol{n} = \boldsymbol{e}_2) = \tau_{yx} \boldsymbol{e}_1 + \sigma_y \boldsymbol{e}_2 + \tau_{yz} \boldsymbol{e}_3 \\ \boldsymbol{T}^n(x, \boldsymbol{n} = \boldsymbol{e}_3) = \tau_{zx} \boldsymbol{e}_1 + \tau_{zy} \boldsymbol{e}_2 + \sigma_z \boldsymbol{e}_3 \end{cases} \tag{3.5}$$

where $\boldsymbol{e}_1$, $\boldsymbol{e}_2$, $\boldsymbol{e}_3$ are the unit vectors along each coordinate direction, and the nine quantities $\{\sigma_x, \sigma_y, \sigma_z, \tau_{xy}, \tau_{yx}, \tau_{yz}, \tau_{zy}, \tau_{zx}, \tau_{xz}\}$ are the components of the traction vector on each of three coordinate planes as illustrated. These nine components are called the *stress components*, with σ_x, σ_y, σ_z called the *normal stresses* and τ_{xy}, τ_{yx}, τ_{yz}, τ_{zy}, τ_{zx}, τ_{xz} called the *shearing stresses*. The components of stress σ_{ij} are commonly written in matrix format

$$\boldsymbol{\sigma} = [\sigma] = \begin{bmatrix} \sigma_x & \tau_{xy} & \tau_{xz} \\ \tau_{yx} & \sigma_y & \tau_{yz} \\ \tau_{zx} & \tau_{zy} & \sigma_z \end{bmatrix} \tag{3.6}$$

and it can be formally seen that the stress is a second-order tensor that obeys the appropriate transformation law.

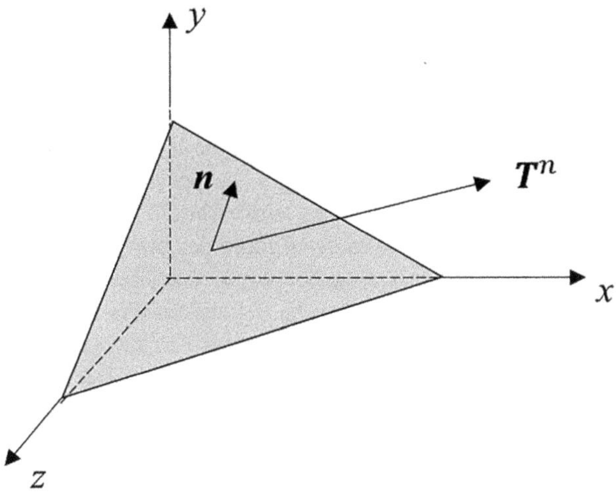

Fig. 3.4 Traction on an oblique plane

The positive directions of each stress component are illustrated in Fig. 3.3. Regardless of the coordinate system, positive normal stress always acts in tension out of the face, and only one subscript is necessary because it always acts normal to the surface. The shear stress, however, requires two subscripts, the first representing the plane of action and the second designating the direction of the stress. Similar to shear strain, the sign of the shear stress depends on the orientation of the coordinate system. For example, on a plane with a normal in the positive x direction, positive τ_{xy} acts in the positive y direction. Similar definitions follow for other shear stress components. In the subsequent chapters, proper formulation of elasticity problems requires knowledge of these basic definitions, directions, and sign conventions for particular stress components.

Next, the traction vector on an oblique plane with arbitrary orientation, as shown in Fig. 3.4, is considered. The unit normal to the surface can be expressed as

$$\boldsymbol{n} = n_x \boldsymbol{e}_1 + n_y \boldsymbol{e}_2 + n_2 \boldsymbol{e}_3 \tag{3.7}$$

where n_x, n_y, n_z are the direction cosines of the unit vector $\boldsymbol{n}$ relative to the given coordinate system. We now consider the equilibrium of the pyramidal element interior to the oblique and coordinate planes. Invoking the force balance between tractions on the oblique and coordinate faces gives

$$\boldsymbol{T}^n = n_x \boldsymbol{T}^n(\boldsymbol{n} = \boldsymbol{e}_1) + n_y \boldsymbol{T}^n(\boldsymbol{n} = \boldsymbol{e}_2) + n_z \boldsymbol{T}^n(\boldsymbol{n} = \boldsymbol{e}_3) \tag{3.8}$$

and using Eq. (3.5), this can be written as

$$\begin{aligned} \boldsymbol{T}^n = & \left(\sigma_x n_x + \tau_{yx} n_y + \tau_{zx} n_z\right)\boldsymbol{e}_1 \\ & + \left(\tau_{xy} n_x + \sigma_y n_y + \tau_{zy} n_z\right)\boldsymbol{e}_2 \\ & + \left(\tau_{xz} n_x + \tau_{yz} n_y + \sigma_z n_z\right)\boldsymbol{e}_3 \end{aligned} \tag{3.9}$$

or in index notation

$$T_i^n = \sigma_{ji} n_j \tag{3.10}$$

Equation (3.9) or Eq. (3.10) provides a simple and direct method for calculating the forces on oblique planes and surfaces. This technique proves to be very useful for specifying general boundary conditions during the formulation and solution of elasticity problems. Based on these previous definitions, the distinction between the traction vector and stress tensor should be carefully understood. Although each quantity has the same units of force per unit area, they are fundamentally different since the traction is a vector while the stress is a second-order tensor (matrix). Components of traction can be defined on any surface, but particular stress components only exist on coordinate surfaces, as shown in Fig. 3.3 for the Cartesian case. Clearly, Eq. (3.10) establishes the relation between the two variables, thereby indicating that each traction component can be expressed as a linear combination of particular stress components.

Following the principles of small deformation theory, the previous definitions for the stress tensor and traction vector do not make a distinction between the deformed and undeformed configurations of the body. As mentioned in the previous chapter, such a distinction only leads to small modifications that are considered higher-order effects and are normally neglected. However, for the large deformation theory, sizeable differences exist between these configurations, and the undeformed configuration (commonly called the *reference configuration*) is often used in problem formulation. This gives rise to the definition of an additional stress called the *Piola-Kirchhoff stress tensor* that represents the force per unit area in the reference configuration. In the more general scheme, the stress σ_{ij} is referred to as the *Cauchy stress tensor*. Throughout this textbook only small deformation theory is considered, and thus the distinction between these two definitions of stress does not exist, thereby any need for this additional terminology is eliminated.

$$\begin{gathered} N = \boldsymbol{T}^n \cdot \boldsymbol{n} \\ S = \left(|\boldsymbol{T}^n|^2 - N^2 \right)^{1/2} \end{gathered} \tag{3.11}$$

where N and S are the normal and shear stresses at the sectioned surface,

If the Cartesian planes are taken as the projected planes (Fig. 3.5), i.e. x-y, x-z and y-z planes, the traction vector can be expressed using the stress components as below:

$$\boldsymbol{f}_v = f_{vx}\boldsymbol{e}_1 + f_{vy}\boldsymbol{e}_2 + f_{vz}\boldsymbol{e}_3 \tag{3.12}$$

where, f_{vx} and f_{vy} are the stress components in the x-y plane along the x and y axes, respectively, f_{vz} is the stress component in the x-z plane along the z axis, $\boldsymbol{e}_1$, $\boldsymbol{e}_2$, $\boldsymbol{e}_3$ are the basis vectors along the x, y and z axes, respectively.

It is revealed in Fig. 3.5 that if the traction vector is projected along the three orthogonal Cartesian planes, three stress components can be obtained for each section. If the section planes are parallel to the Cartesian planes, nine stress

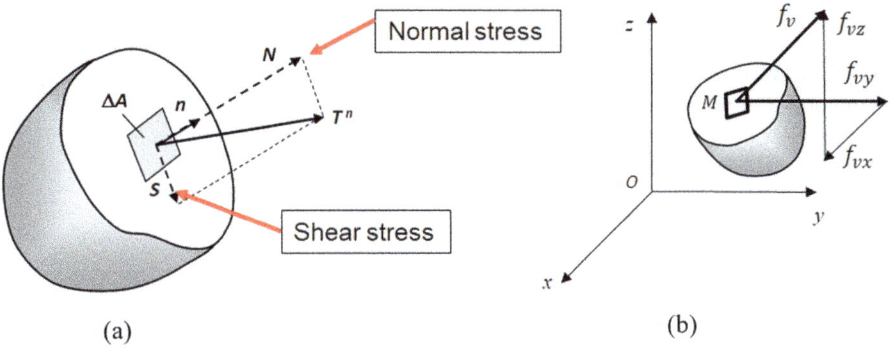

Fig. 3.5 Projections of the traction vector into (**a**) normal and shear stresses; (**b**) stress components along the x, y and z axes

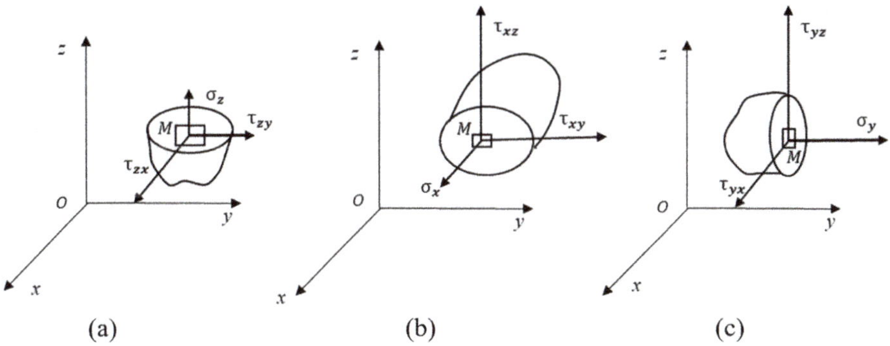

Fig. 3.6 Projections of the traction vector in the (**a**) x-y, (**b**) y-z and (**c**) x-z planes

components are obtained (Fig. 3.6). The nine components form the stress tensor and are used to express the stress at a point within the deformable body.

It should be noted that because

$$\tau_{xy} = \tau_{yx}, \tau_{xz} = \tau_{zx}, \tau_{zy} = \tau_{yz} \tag{3.13}$$

the stress tensor is a symmetric matrix. To save the storage space and accelerate the calculation in the engineering computation, the kinematic rule can be applied to simplify the stress tensor to a six-component vector as below.

$$\boldsymbol{\sigma} = \begin{bmatrix} \sigma_x & \sigma_y & \sigma_z & \tau_{yz} & \tau_{xz} & \tau_{xy} \end{bmatrix}^{\mathrm{T}} \tag{3.14}$$

It also should be noted that in some engineering textbooks, the following six-component vector is used:

$$\boldsymbol{\sigma} = \begin{bmatrix} \sigma_x & \sigma_y & \sigma_z & \tau_{xy} & \tau_{xz} & \tau_{yz} \end{bmatrix}^{\mathrm{T}} \tag{3.15}$$

Both expressions (Eqs. (3.14) and (3.15)) are correct. As long as the readers or engineers use one of them consistently in their calculations, the results will not be affected. Another remark is that once one form of the stress vector is used, the corresponding strain vector should also be used accordingly.

3.2.3 Type of Stress

From the definitions of stress and strain, it can be seen the stress is the cause of a deformation and the strain is the consequence of the deformation. Depend on the force state of the deformable body, the stress can be classified into the tensile stress and the compressive stress. When equal and opposite external forces are directed away from each other (Fig. 3.7a), the stress generated inside the deformable body is the tensile stress. In contrast, when equal and opposite external forces are directed toward each other (Fig. 3.7b), the stress generated inside is the compressive stress.

In addition, the six components of the stress tensor can be classified into normal stresses and shear stresses. The normal stresses, i.e., σ_x, σ_y, σ_z, are the ones causing the stretching or shortening deformations of the deformable body (Fig. 3.8). The normal stresses are acting in the normal direction of the differential volume (Fig. 3.8). The direction of the normal stress is defined based on the deformation state of the deformable body. If the normal stress causes the stretching of the body, the normal stress is positive and sometimes it is called the *tensile normal stress*. In contrast, if the normal stress causes the shortening of the body, the normal stress is negative and it can also be called the *compressive normal stress*.

The shear stresses, i.e., τ_{xy}, τ_{yz}, τ_{zx} are the ones causing the shape change of the deformable body (Fig. 3.9). The shear stresses are acting on the tangent direction of the differential volume (Fig. 3.9) and their directions are defined based on both the direction of the Cartesian coordinate axis and the outer normal direction of the differential volume. Under the following two cases, the shear stresses are positive:

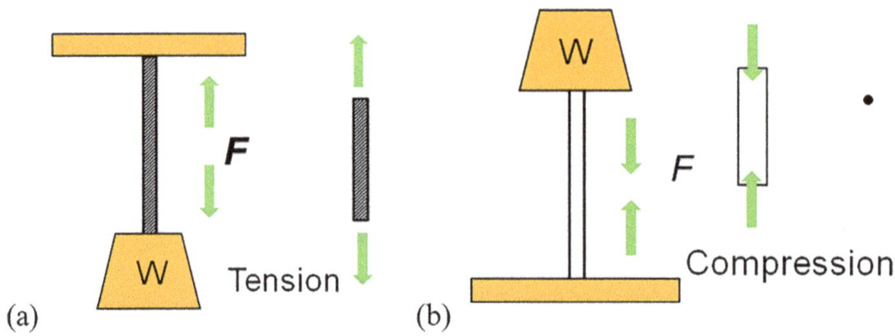

Fig. 3.7 Classification of stress based on the force state: (**a**) tension; (**b**) compression

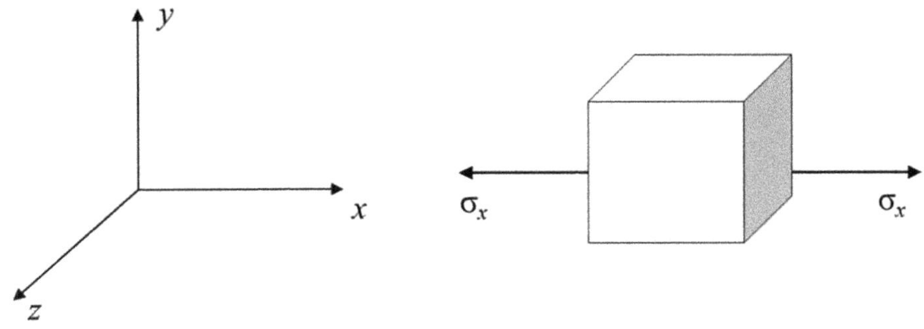

Fig. 3.8 Illustration of the normal stress

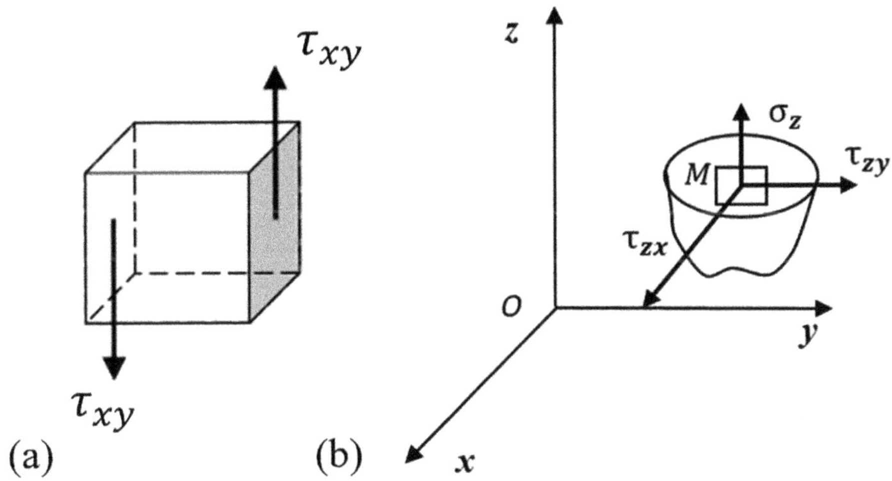

Fig. 3.9 Illustration of the shear stress components: (**a**) τ_{xy}, (**b**) τ_{zx} and τ_{zy}

(1) When the outer normal direction of the plane, where the shear stress is on, is along the positive direction of the Cartesian axis and the shear stress is along the positive direction of the axis. For example, in Fig. 3.9b, τ_{zx} is positive, because the outer normal direction of the section plane is along the positive direction of z direction and in the section plane, τ_{zx} is along the positive direction of x axis.
(2) When the outer normal direction of the plane, where the shear stress is on, is along the negative direction of the Cartesian axis and the shear stress is along the negative direction of the axis.

It can be seen when both the outer normal direction of the section plane and the axis direction of the shear stress are consistent, i.e., both are positive or negative, and then the shear stress is positive. When outer normal direction of the section plane and the axis direction of the shear stress are not consistent, i.e., one positive and the other negative, and then the shear stress is negative. A summary of the justification of the direction of the shear stress is presented in Table 3.1.

Table 3.1 Justification of the direction of the shear stress

Whether the outer normal direction of the section plane is along the Cartesian axis	Whether the shear stress is along the Cartesian axis	Direction of the shear stress
Yes	Yes	Positive
No	No	Positive
Yes	No	Negative
No	Yes	Negative

3.3 Stress Transformation

3.3.1 *Stress Transformation Using the Tensor Operation*

Analogous to our previous discussion with the strain tensor, the stress components must also follow the standard transformation rules for second-order tensors. Applying transformation relation for the stress gives

$$\sigma'_{ij} = Q_{ip}Q_{jq}\sigma_{pq} \tag{3.16}$$

where $Q_{ij} = \cos\left(x'_i, x_j\right)$ is the rotation matrix. Therefore, given the stress in one coordinate system, we can determine the new components in any other rotated system. For the general 3D case, the rotation matrix may be chosen in the form

$$Q_{ij} = \begin{bmatrix} l_1 & m_1 & n_1 \\ l_2 & m_2 & n_2 \\ l_3 & m_3 & n_3 \end{bmatrix} \tag{3.17}$$

Using this notational scheme, the specific transformation relations for the stress then become

$$\begin{cases} \sigma'_x = \sigma_x l_1^2 + \sigma_y m_1^2 + \sigma_z n_1^2 + 2\left(\tau_{xy} l_1 m_1 + \tau_{yz} m_1 n_1 + \tau_{zx} n_1 l_1\right) \\ \sigma'_y = \sigma_x l_2^2 + \sigma_y m_2^2 + \sigma_z n_2^2 + 2\left(\tau_{xy} l_2 m_2 + \tau_{yz} m_2 n_2 + \tau_{zx} n_2 l_2\right) \\ \sigma'_z = \sigma_x l_3^2 + \sigma_y m_3^2 + \sigma_z n_3^2 + 2\left(\tau_{xy} l_3 m_3 + \tau_{yz} m_3 n_3 + \tau_{zx} n_3 l_3\right) \\ \tau'_{xy} = \sigma_x l_1 l_2 + \sigma_y m_1 m_2 + \sigma_z n_1 n_2 + \tau_{xy}(l_1 m_2 + m_1 l_2) + \tau_{yz}(m_1 n_2 + n_1 m_2) \\ \qquad + \tau_{zx}(n_1 l_2 + l_1 n_2) \\ \tau'_{yz} = \sigma_x l_2 l_3 + \sigma_y m_2 m_3 + \sigma_z n_2 n_3 + \tau_{xy}(l_2 m_3 + m_2 l_3) + \tau_{yz}(m_2 n_3 + n_2 m_3) \\ \qquad + \tau_{zx}(n_2 l_3 + l_2 n_3) \\ \tau'_{zx} = \sigma_x l_3 l_1 + \sigma_y m_3 m_1 + \sigma_z n_3 n_1 + \tau_{xy}(l_3 m_1 + m_3 l_1) + \tau_{yz}(m_3 n_1 + n_3 m_1) \\ \qquad + \tau_{zx}(n_3 l_1 + l_3 n_1) \end{cases} \tag{3.18}$$

For 2D case, the in-plane stress components can be transformed according to

$$\begin{cases} \sigma_x' = \sigma_x \cos^2\theta + \sigma_y \sin^2\theta + 2\tau_{xy} \sin\theta \cos\theta \\ \sigma_y' = \sigma_x \sin^2\theta + \sigma_y \cos^2\theta - 2\tau_{xy} \sin\theta \cos\theta \\ \tau_{xy}' = -\sigma_x \sin\theta \cos\theta + \sigma_y \sin\theta \cos\theta + \tau_{xy}(\cos^2\theta - \sin^2\theta) \end{cases} \tag{3.19}$$

which is commonly rewritten in the double angle format

$$\begin{cases} \sigma_x' = \dfrac{\sigma_x + \sigma_y}{2} + \dfrac{\sigma_x - \sigma_y}{2} \cos 2\theta + \tau_{xy} \sin 2\theta \\ \sigma_y' = \dfrac{\sigma_x + \sigma_y}{2} - \dfrac{\sigma_x - \sigma_y}{2} \cos 2\theta - \tau_{xy} \sin 2\theta \\ \tau_{xy}' = \dfrac{\sigma_y - \sigma_x}{2} \sin 2\theta + \tau_{xy} \cos 2\theta \end{cases} \tag{3.20}$$

Similar to our discussion on strain in the previous chapter, Eq. (3.20) can be directly applied to establish stress transformations between Cartesian and Polar coordinate systems.

3.3.2 *Stress Transformation Using the Equilibrium Conditions*

The stress transformation can also be used to work out the stress tensor on a 2D oblique plane. To illustrate the calculation process, a differential triangular shown in Fig. 3.10 is used. The length and height of the differential triangular are *dx* and *dy*, respectively. The thickness of the triangular is *t*. The stress tensor at point O can be written as:

$$\boldsymbol{\sigma} = \begin{bmatrix} \sigma_x & \tau_{xy} \\ \tau_{yx} & \sigma_y \end{bmatrix} \tag{3.21}$$

At an oblique plane *AB*, the traction vector f_v can be projected into f_{vx} and f_{vy} along the *x* and *y* directions, respectively. It should be noted the triangular *OAB* is a

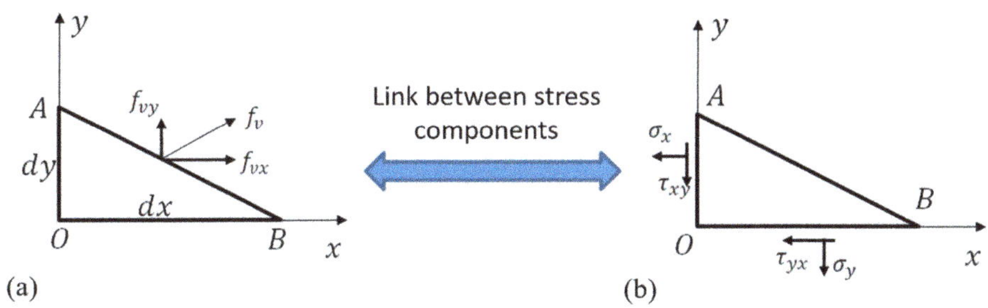

Fig. 3.10 Calculation of stress on a 2D oblique plane: (**a**) stress components in the oblique plane *AB*; (**b**) stress components in the orthogonal planes

differential area, and thus the line segment AB is very close to the point O. In addition, the differential triangular is in the balanced state. Therefore, relation between the four stress components and f_{vx}, f_{vy} can be obtained using the equilibrium equation:

First, the summation of all the forces in the x direction is equal to zero, i.e., $\sum F_x = 0$

$$f_{vx}\Delta S_{AB} - \sigma_x \Delta S_{AO} - \tau_{yx}\Delta S_{OB} - \frac{1}{3}\Delta S_{OAB}\Delta h F_x = 0 \tag{3.22}$$

where ΔS_{AB}, ΔS_{AO}, ΔS_{OB} and ΔS_{OAB} are the areas, and F_x is the body force.

Similarly, $\sum F_y = 0$

$$f_{vy}\Delta S_{AB} - \sigma_y \Delta S_{OB} - \tau_{xy}\Delta S_{OA} - \frac{1}{3}\Delta S_{OAB}\Delta h F_y = 0 \tag{3.23}$$

The higher order term, i.e., $\frac{1}{3}\Delta S_{OAB}\Delta h F_x$, in the equations can be ignored. Then the following relations can be obtained:

$$\begin{cases} f_{vx} = \sigma_x l + \tau_{yx} m \\ f_{vy} = \tau_{xy} l + \sigma_y m \end{cases} \tag{3.24}$$

where, l and m are the cosines between the normal direction and the orthogonal planes, i.e.,

$$\begin{cases} l = \dfrac{\Delta S_{OA}}{\Delta S_{AB}} = cos\alpha \\ m = \dfrac{\Delta S_{OB}}{\Delta S_{AB}} = cos\beta \end{cases} \tag{3.25}$$

Now we can move to the calculation of the traction on the 3D oblique plane using the 9 stress components. The differential tetrahedron in Fig. 3.11 is used to illustrate the calculation process. As discussed previously, the stress tensor at point M can be expressed using the stress components at the three orthogonal differential planes (Fig. 3.11a). At the oblique differential plane (Fig. 3.11b), the traction at any point can be written as $T = (f_{vx}, f_{vy}, f_{vz})$, where f_{vx}, f_{vy}, f_{vz} are the projections along the x, y and z axes. The equilibrium conditions can be used to establish the relationship between the 9 stress components and the traction vector at the oblique plane, i.e., to calculate the stress at the oblique plane using the stress components. Because the tetrahedron is a small differential volume, the oblique differential plane abc can be regarded as the plane very close to the point M. Therefore, first, from the condition $\sum F_x = 0$, the following relation can be set:

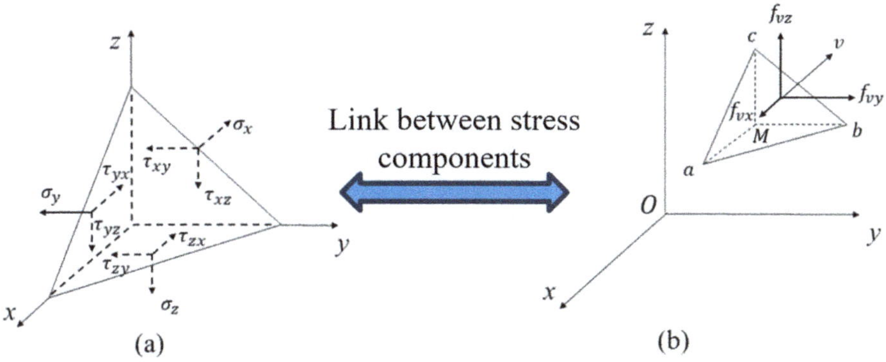

Fig. 3.11 Calculation of stress on a 3D oblique plane: (**a**) 9 stress components in the orthogonal planes; (**b**) 3 stress components in the oblique plane

$$f_{vx}\Delta S_{abc} - \sigma_x \Delta S_{bMc} - \tau_{yx}\Delta S_{aMc} - \tau_{zx}\Delta S_{aMb} + \frac{1}{3}\Delta S_{abc}\Delta h F_x = 0 \tag{3.26}$$

where Δh is the distance from M to the oblique plane abc, F_x is the body force, ΔS_{abc}, ΔS_{bMc}, ΔS_{aMc} and ΔS_{abc} are the differential area of the planes. From Fig. 3.11b, it can be seen the following relations are satisfied:

$$\begin{cases} \Delta S_{bMc} = \Delta S_{abc} l \\ \Delta S_{aMc} = \Delta S_{abc} m \\ \Delta S_{aMb} = \Delta S_{abc} n \end{cases} \tag{3.27}$$

where l, m, n are the cosines between the three orthogonal planes and the oblique plane. Putting the relations in Eq. (3.27) back into Eq. (3.26) and ignoring the higher order part, i.e., $\frac{1}{3}\Delta S_{abc}\Delta h F_x$, we can obtain

$$f_{vx} = \sigma_x l + \tau_{yx} m + \tau_{zx} n \tag{3.28}$$

Similarly, from $\sum F_y = 0$, we can obtain

$$f_{vy} = \tau_{xy} l + \sigma_y m + \tau_{zy} n \tag{3.29}$$

From $\sum F_z = 0$, we can obtain

$$f_{vz} = \tau_{xz} l + \tau_{yx} m + \sigma_z n \tag{3.30}$$

Equations (3.28)–(3.30) are used to work out the traction vector at the oblique plane from the nine stress components.

3.3.3 Stress Transformation Induced by the Change of Coordinate System

In the previous section, the formulations for calculating the traction at the oblique plane from the nine stress components have been given. In this part, the stress transformation induced by the change of the coordinate system is explored. The change of coordinate system consists of two cases. The first case is the translation of the coordinate system. In this case, the nine stress components stay the same. The second case is the rotation of the coordinate system. As shown in Fig. 3.12, the $Oxyz$ coordinate system is rotated to a new system $Ox'y'z'$. In the new system $Ox'y'z'$, the stress tensor will be different but can be calculated from that in the old coordinate. The stress tensor in the old system is:

$$(\sigma_{ij}) = \begin{bmatrix} \sigma_x & \tau_{xy} & \tau_{xz} \\ \tau_{yx} & \sigma_y & \tau_{yz} \\ \tau_{zx} & \tau_{zy} & \sigma_z \end{bmatrix} \tag{3.31}$$

To work out the stress tensor in the $Ox'y'z'$ system, the cosine values from the axes of the new system to the axes of the old system are worked out and present in Table 3.2.

In the table above, l_i, m_i, n_i ($i = 1, 2, 3$) are the cosines from the axis of the new system to the axis of the old system.

The stress tensor in the $Ox'y'z'$ system is written as:

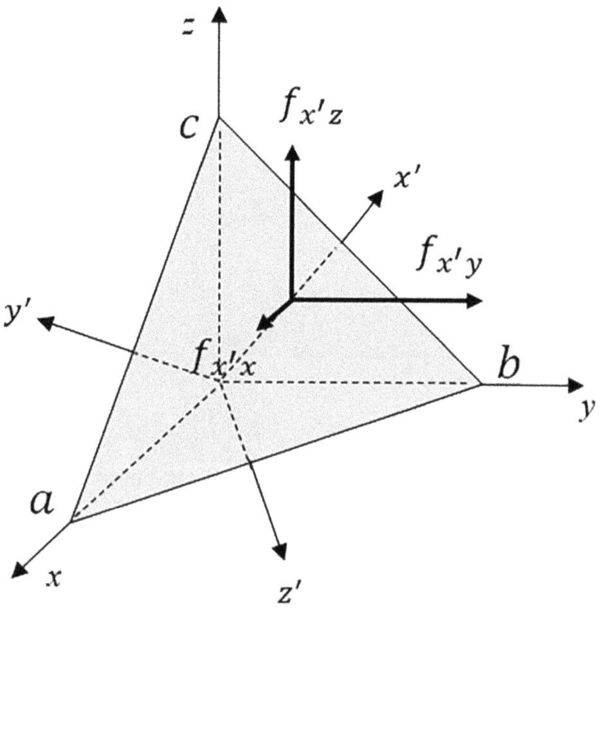

Fig. 3.12 Change of the coordinate system

Table 3.2 Relation between the ***Oxyz*** and ***Ox′y′z′*** systems

	x	y	z
x'	l_1	m_1	n_1
y'	l_2	m_2	n_2
z'	l_3	m_3	n_3

$$\left(\sigma_{i'j'}\right) = \begin{bmatrix} \sigma_{x'} & \tau_{x'y'} & \tau_{x'z'} \\ \tau_{y'x'} & \sigma_{y'} & \tau_{y'z'} \\ \tau_{z'x'} & \tau_{z'y'} & \sigma_{z'} \end{bmatrix} \tag{3.32}$$

where $\sigma_{x'}$, $\tau_{x'y'}$, $\tau_{x'z'}$ are the components in the differential plane perpendicular to ox' axis along the ox', oy' and oz' axes, respectively. Similarly, $\sigma_{y'}$, $\tau_{y'x'}$, $\tau_{y'z'}$ are the stress components in the differential plane perpendicular to Oy' axis, and $\sigma_{z'}$, $\tau_{z'x'}$, $\tau_{z'y'}$ are the stress components in the differential plane perpendicular to Oz' axis. $\sigma_{x'}$, $\tau_{x'y'}$, $\tau_{x'z'}$ are selected as the variables to demonstrate the calculation process. In the oblique plane, the traction vector is projected along the Ox', Oy' and Oz' axes and $f_{x'x}$, $f_{x'y}$ and $f_{x'z}$ are the corresponding components. Then the following relations can be obtained:

$$\begin{cases} \sigma_{x'} = \boldsymbol{f}_{x'} \cdot \boldsymbol{e}_1' = f_{x'x} l_1 + f_{x'y} m_1 + f_{x'z} n_1 \\ \tau_{x'y'} = \boldsymbol{f}_{x'} \cdot \boldsymbol{e}_2' = f_{x'x} l_2 + f_{x'y} m_2 + f_{x'z} n_2 \\ \tau_{x'z'} = \boldsymbol{f}_{x'} \cdot \boldsymbol{e}_3' = f_{x'x} l_3 + f_{x'y} m_3 + f_{x'z} n_3 \end{cases} \tag{3.33}$$

where, $\boldsymbol{e}_1'$, $\boldsymbol{e}_2'$ and $\boldsymbol{e}_3'$ are the unit directional vectors in the Ox', Oy' and Oz' axes, $\boldsymbol{f}_{x'}$ is force vector in the Ox' axis. In addition, $f_{x'x}$,$f_{x'y}$ and $f_{x'z}$ are the components of the traction in the oblique plane with respect to the old system. Their relations with the 9 stress components can be written as:

$$\begin{cases} f_{x'x} = \sigma_x l_1 + \tau_{yx} m_1 + \tau_{zx} n_1 \\ f_{x'y} = \tau_{xy} l_1 + \sigma_y m_1 + \tau_{zy} n_1 \\ f_{x'z} = \tau_{xz} l_1 + \tau_{yz} m_1 + \sigma_z n_1 \end{cases} \tag{3.34}$$

Putting Eq. (3.34) back to Eq. (3.33), the relation between $\sigma_{x'}$, $\tau_{x'y'}$, $\tau_{x'z'}$ and the nine stress components in the old system can be obtained as:

$$\begin{cases} \sigma_{x'} = \sigma_x l_1^2 + \sigma_y m_1^2 + \sigma_z n_1^2 + 2\tau_{yz} m_1 n_1 + 2\tau_{zx} l_1 n_1 + 2\tau_{xy} l_1 m_1 \\ \tau_{x'y'} = \sigma_x l_1 l_2 + \sigma_y m_1 m_2 + \sigma_z n_1 n_2 + \tau_{yz}(m_1 n_2 + m_2 n_1) \\ \qquad + \tau_{xz}(l_1 n_2 + l_2 n_1) + \tau_{xy}(l_1 m_2 + l_2 m_1) \\ \tau_{x'z'} = \sigma_x l_1 l_3 + \sigma_y m_1 m_3 + \sigma_z n_1 n_3 + \tau_{yz}(m_1 n_3 + m_3 n_1) \\ \qquad + \tau_{xz}(l_1 n_3 + l_3 n_1) + \tau_{xy}(l_1 m_3 + l_3 m_1) \end{cases} \tag{3.35}$$

Following the similar procedure, the relation between $\sigma_{y'}$, $\tau_{y'x'}$, $\tau_{y'z'}$ and the nine stress components in the old system can be obtained as:

$$\begin{cases} \sigma_{y'} = \sigma_x l_2^2 + \sigma_y m_2^2 + \sigma_z n_2^2 + 2(\tau_{xy} l_2 m_2 + \tau_{yz} m_2 n_2 + \tau_{zx} n_2 l_2) \\ \tau_{y'x'} = \sigma_x l_1 l_2 + \sigma_y m_1 m_2 + \sigma_z n_1 n_2 + \tau_{yz}(m_1 n_2 + m_2 n_1) \\ \qquad + \tau_{xz}(l_1 n_2 + l_2 n_1) + \tau_{xy}(l_1 m_2 + l_2 m_1) \\ \tau_{y'z'} = \sigma_x l_2 l_3 + \sigma_y m_2 m_3 + \sigma_z n_2 n_3 + \tau_{xy}(l_2 m_3 + m_2 l_3) \\ \qquad + \tau_{yz}(m_2 n_3 + n_2 m_3) + \tau_{zx}(n_2 l_3 + l_2 n_3) \end{cases} \tag{3.36}$$

The relation between $\sigma_{z'}$, $\tau_{z'x'}$, $\tau_{z'y'}$ and the nine stress components in the old system can be obtained as:

$$\begin{cases} \sigma_{z'} = \sigma_x l_3^2 + \sigma_y m_3^2 + \sigma_z n_3^2 + 2(\tau_{xy} l_3 m_3 + \tau_{yz} m_3 n_3 + \tau_{zx} n_3 l_3) \\ \tau_{z'x'} = \sigma_x l_3 l_1 + \sigma_y m_3 m_1 + \sigma_z n_3 n_1 + \tau_{xy}(l_3 m_1 + m_3 l_1) \\ \qquad + \tau_{yz}(m_3 n_1 + n_3 m_1) + \tau_{zx}(n_3 l_1 + l_3 n_1) \\ \tau_{z'y'} = \sigma_x l_2 l_3 + \sigma_y m_2 m_3 + \sigma_z n_2 n_3 + \tau_{xy}(l_2 m_3 + m_2 l_3) \\ \qquad + \tau_{yz}(m_2 n_3 + n_2 m_3) + \tau_{zx}(n_2 l_3 + l_2 n_3) \end{cases} \tag{3.37}$$

Now let's use the 2D case to further demonstrate the stress transformation induced by the rotation of coordinate system. As shown in Fig. 3.13, a 2D coordinate system *Oxy* is rotated by φ and the new system $Ox'y'$ is formed. The cosines from the axes in the *Oxy* system to the axes in the $Ox'y'$ systems are presented in Table 3.3.

Table 3.3 can be reduced to Table 3.4 shown below:

Then, based on Eq. (3.16), we have:

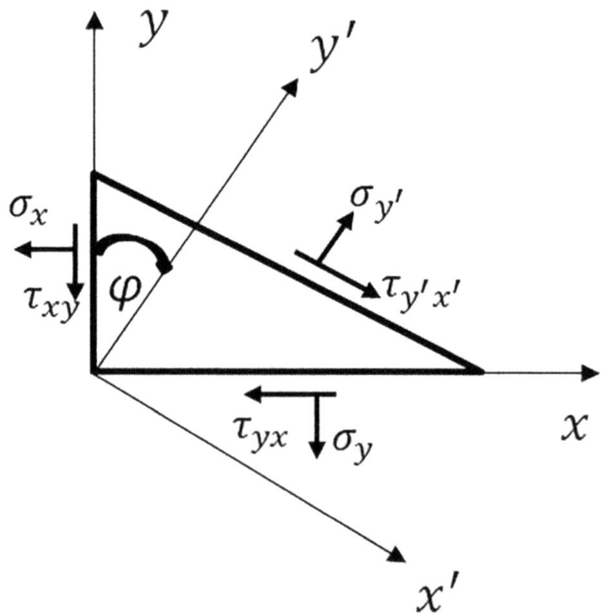

Fig. 3.13 Stress transformation induced by rotation of coordinate system in 2D case

Table 3.3 Relation between the ***Oxy*** and ***Ox′y′*** systems

	x	y
x'	cos (−φ)	cos[−(90° + φ)]
y'	cos(90° − φ)	cos(−φ)

Table 3.4 Relation between the ***Oxy*** and ***Ox′y′*** systems

	x	y
x'	cosφ	−sinφ
y'	sinφ	cosφ

$$\begin{cases} \sigma_{x'} = \sigma_x l_1^2 + \sigma_y m_1^2 + 2\tau_{xy} l_1 m_1 \\ \sigma_{y'} = \sigma_x l_2^2 + \sigma_y m_2^2 + 2\tau_{xy} l_2 m_2 \\ \tau_{x'y'} = \sigma_x l_1 l_2 + \sigma_y m_1 m_2 + \tau_{xy}(l_1 m_2 + l_2 m_1) \end{cases} \tag{3.38}$$

Putting the values in Table 3.4 into the relations above, we have:

$$\begin{cases} \sigma_{x'} = \sigma_x \cos^2\varphi + \sigma_y \sin^2\varphi - 2\tau_{xy} \cos\varphi\sin\varphi \\ \sigma_{y'} = \sigma_x \sin^2\varphi + \sigma_y \cos^2\varphi + 2\tau_{xy} \cos\varphi\sin\varphi \\ \tau_{x'y'} = (\sigma_x - \sigma_y) \cos\varphi\sin\varphi + \tau_{xy}(\cos^2\varphi - \sin^2\varphi) \end{cases} \tag{3.39}$$

which is the relation between the stress components in the new coordinate system with those in the old coordinate system.

In addition to the method presented above to perform the stress transformation, the transformation can also be done by using the equilibrium equations as below. Figure 3.14 is used to illustrate the calculation process. The coordinate system *Oxy* is rotated anti-clockwise by an angle of θ and then the new coordinate system $Ox'y'$ is formed. The triangular *OBC* is regarded as a differential volume with the thickness of 1.0, and the area of *BC* is set to 1.0. Consequently, the areas of *AB* and *AC* are $cos\theta$ and $sin\theta$. In the *Oxy* system, the stress components at the *OB* face are σ_x and τ_{xy}, the stress components at the OC face are σ_y and τ_{yx} and the traction components at the BC face are p_x and p_y. Based on the equilibrium condition, $\sum F_x = 0$, we can obtain:

$$p_x = \sigma_x \cos\theta + \tau_{xy} \sin\theta \tag{3.40}$$

Similarly, from $\sum F_y = 0$, we can obtain:

$$p_y = \tau_{xy} \cos\theta + \sigma_y \sin\theta \tag{3.41}$$

In addition, the traction p_x and p_y can be projected along the Ox' and Oy' axes, and then we have the following relations:

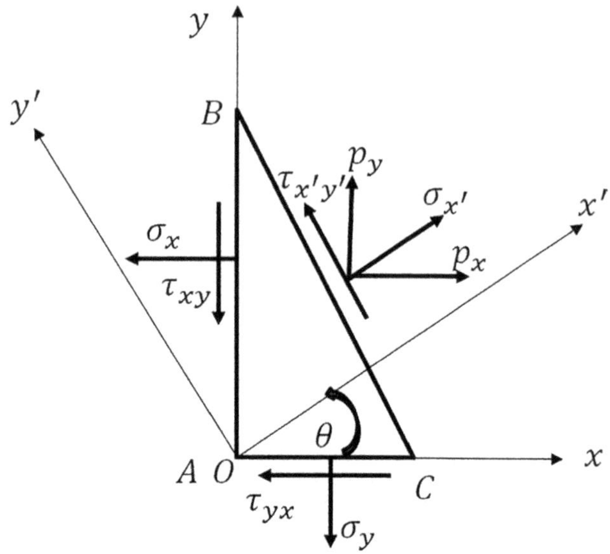

Fig. 3.14 Illustration of the stress transformation induced by the rotation of coordinate system in 2D case

$$\begin{cases} \sigma_{x'} = p_x \cos\theta + p_y \sin\theta \\ \tau_{x'y'} = p_y \cos\theta - p_x \sin\theta \end{cases} \tag{3.42}$$

In Eq. (3.42), the variables p_x and p_y can be eliminated, and then the following relations can be obtained:

$$\begin{cases} \sigma_{x'} = \sigma_x cos^2\theta + \sigma_y sin^2\theta + 2\tau_{xy} sin\,\theta cos\theta \\ \tau_{x'y'} = \tau_{xy}\left(cos^2\theta - sin^2\theta\right) + \left(\sigma_y - \sigma_x\right) sin\,\theta cos\theta \end{cases} \tag{3.43}$$

Similarly, the relation between $\sigma_{y'}$, $\tau_{y'x'}$ and the four stress components in the old coordinate system can be worked out as:

$$\begin{cases} \sigma_{y'} = \sigma_x sin^2\theta + \sigma_y cos^2\theta - 2\tau_{xy} sin\,\theta cos\theta \\ \tau_{x'y'} = \tau_{xy}\left(cos^2\theta - sin^2\theta\right) + \left(\sigma_y - \sigma_x\right) sin\,\theta cos\theta \end{cases} \tag{3.44}$$

Equations (3.43) and (3.44) can be used to transform the stress tensor from the old coordinate system Oxy to the new system $Ox'y'$.

Now let's look at the stress transformation in 3D case. A small differential tetrahedron $OABC$ in Fig. 3.15 is used to illustrate the calculation process. The area of the face ABC is set to 1.0. Then based on the relations in Fig. 3.15, the areas of OBC, OAC and OAB can be worked out as:

$$\text{Area}(OBC) = 1 \times \cos(n, x) = l_1$$

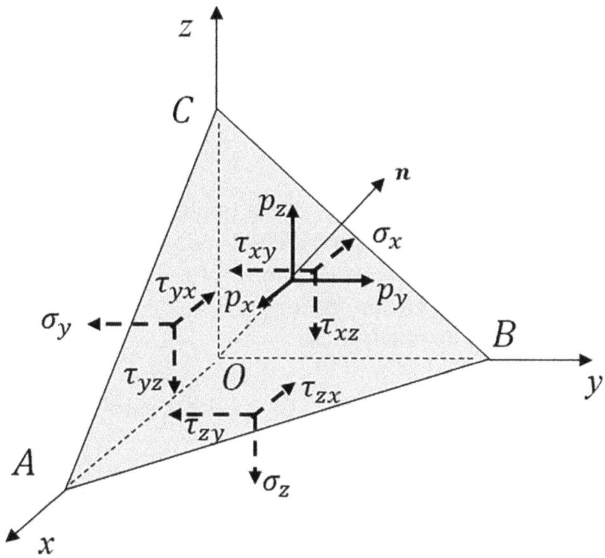

Fig. 3.15 Illustration of the stress transformation induced by the rotation of coordinate system in 3D case

$$\text{Area}(OAC) = 1 \times \cos(n, y) = l_2$$
$$\text{Area}(OAB) = 1 \times \cos(n, z) = l_3$$

where, n represents the normal direction of the face ABC. The nine stress components in the three orthogonal planes are plotted in Fig. 3.15. In the oblique plane ABC, the components of the traction vector are represented as p_x, p_y and p_z. First, based on the equilibrium equation $\sum F_x = 0$, the following relation can be obtained:

$$p_x = \sigma_x l_1 + \tau_{yx} l_2 + \tau_{zx} l_3 \tag{3.45}$$

Similarly, based on the equations $\sum F_y = 0$ and $\sum F_z = 0$, the following relations can be obtained:

$$\begin{cases} p_y = \tau_{xy} l_1 + \sigma_y l_2 + \tau_{zy} l_3 \\ p_z = \tau_{xz} l_1 + \tau_{yz} l_2 + \sigma_z l_3 \end{cases} \tag{3.46}$$

Because $\tau_{xy} = \tau_{yx}$, which can also be written as $\sigma_{xy} = \sigma_{yx}$, Eqs. (3.45) and (3.46) can be rewritten in the index notation as:

$$p_i = \sigma_{ij} n_j \ (i, j = x, y, z) \tag{3.47}$$

The equation above has established the relation between the nine stress components and the traction components in the oblique plane. Now let's further work out the nine stress components in the new coordinate system $Ox'y'z'$. We set the Ox' axis aligned

Table 3.5 Relation between the ***Oxyz*** and ***Ox′y′z′*** systems

	x	y	z
x'	l_{11}	l_{12}	l_{13}
y'	l_{21}	l_{22}	l_{23}
z'	l_{31}	l_{32}	l_{33}

along the normal direction of the oblique plane, and then the relation between the new coordinate system and the old coordinate system is summarized in Table 3.5.

where l_{11} is the cosine from Ox' to Ox axis, i.e., $l_{11} = \cos(x', x)$. Similarly, $l_{12} = \cos(x', y)$, etc. projecting p_x, p_y and p_z to Ox' axis, the relation between $\sigma_{x'}$ and p_x, p_y, p_z can be obtained as:

$$\sigma_{x'} = p_x l_{11} + p_y l_{12} + p_z l_{13} \tag{3.48}$$

Using Eqs. (3.45) and (3.46) to eliminate p_x, p_y and p_z, the relation between $\sigma_{x'}$ and the stress components in the old coordinate system can be obtained as below:

$$\sigma_{x'} = \sigma_x l_{11}^2 + \sigma_y l_{12}^2 + \sigma_z l_{13}^2 + 2\left(\tau_{xy} l_{11} l_{12} + \tau_{yz} l_{12} l_{13} + \tau_{xz} l_{11} l_{13}\right) \tag{3.49}$$

Similarly, we can obtain

$$\begin{cases} \sigma_{y'} = \sigma_x l_{21}^2 + \sigma_y l_{22}^2 + \sigma_z l_{23}^2 + 2\left(\tau_{xy} l_{21} l_{22} + \tau_{yz} l_{22} l_{23} + \tau_{xz} l_{23} l_{21}\right) \\ \sigma_{z'} = \sigma_x l_{31}^2 + \sigma_y l_{32}^2 + \sigma_z l_{33}^2 + 2\left(\tau_{xy} l_{31} l_{32} + \tau_{yz} l_{32} l_{33} + \tau_{xz} l_{33} l_{31}\right) \\ \sigma_{x'y'} = \sigma_x l_{11} l_{21} + \sigma_y l_{12} l_{22} + \sigma_z l_{13} l_{23} + \tau_{yz}(l_{12} l_{23} + l_{22} l_{13}) + \tau_{xz}(l_{11} l_{23} + l_{21} l_{13}) \\ \qquad + \tau_{xy}(l_{11} l_{22} + l_{21} l_{12}) \\ \sigma_{z'y'} = \sigma_x l_{31} l_{21} + \sigma_y l_{32} l_{22} + \sigma_z l_{33} l_{23} + \tau_{yz}(l_{32} l_{21} + l_{22} l_{33}) + \tau_{xz}(l_{31} l_{23} + l_{21} l_{33}) \\ \qquad + \tau_{xy}(l_{31} l_{22} + l_{21} l_{32}) \\ \sigma_{x'z'} = \sigma_x l_{11} l_{31} + \sigma_y l_{12} l_{32} + \sigma_z l_{13} l_{33} + \tau_{yz}(l_{12} l_{33} + l_{32} l_{13}) + \tau_{xz}(l_{11} l_{33} + l_{31} l_{13}) \\ \qquad + \tau_{xy}(l_{11} l_{32} + l_{31} l_{12}) \end{cases}$$

The equations above can be rewritten in terms of the index notation as:

$$\sigma_{i'j'} = l_{i'i} l_{j'j} \sigma_{ij} \tag{3.50}$$

The equation above has established the relation between the nine stress components in the old coordinate system with those in the rotated (new) coordinate system.

3.4 Equilibrium Equations

The stress field in an elastic solid is continuously distributed within the body and uniquely determined from the applied loadings. Because we are dealing primarily with bodies in equilibrium, the applied loadings satisfy the equations of static

Fig. 3.16 Body and surface forces acting on an arbitrary porting of a continuum

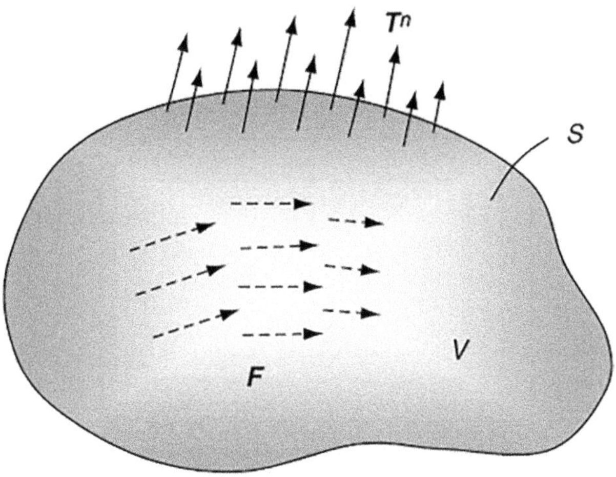

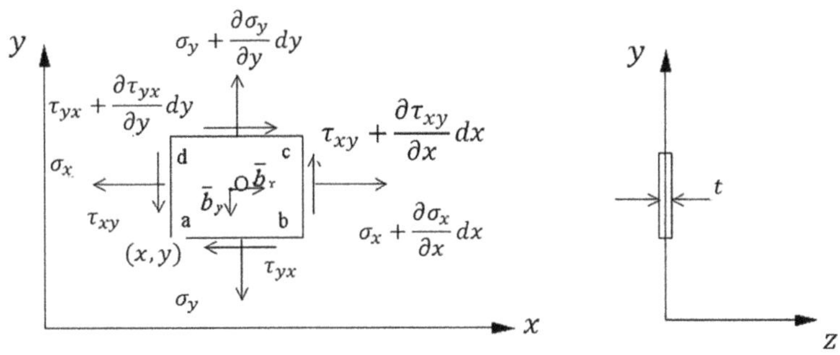

Fig. 3.17 The balanced 2D differential volume

equilibrium; that is, the summation of forces and moments is zero. If the entire body is in equilibrium, then all parts must also be in equilibrium. Thus, we can partition any solid into an appropriate subdomain and apply the equilibrium principle to that region. Following this approach, equilibrium equations can be developed that express the vanishing of the resultant force and moment at a continuum point in the material. These equations can be developed by using either an arbitrary finite subdomain or a special differential region with boundaries coinciding with coordinate surfaces.

A closed subdomain with volume V and surface S within a body in equilibrium is considered. The region has a general distribution of surface tractions $\boldsymbol{T}^n$ and body forces $\boldsymbol{F}$ as shown in Fig. 3.16. For static equilibrium, conservation of linear momentum implies that the forces acting on this region are balanced and thus the resultant force must vanish. This concept can be easily written in index notation as follows.

From the 3D balanced deformable body, a 2D differential volume (Fig. 3.17) is taken as an example to derive the equilibrium equation for 2D case.

The 2D differential volume is balanced and thus the following equilibrium conditions should be satisfied:

(1) Summation of all the forces along the x direction is equal to zero;
(2) Summation of all the forces along the y direction is equal to zero;
(3) Summation of all the moments along any point within the body is equal to zero.

First, from the condition $\sum F_x = 0$, the following equation can be obtained:

$$\begin{gathered}\left(\sigma_x + \frac{\partial \sigma_x}{\partial x}dx\right)dy \cdot t - \sigma_x dy \cdot t + \left(\tau_{yx} + \frac{\partial \tau_{yx}}{\partial y}dy\right)dx \cdot t \\ - \tau_{yx} dx \cdot t + \bar{b}_x dxdy \cdot t = 0\end{gathered} \tag{3.51}$$

where, t is the thickness, $\bar{b}_x$ is the body force.

Through the mathematical operations, Eq. (3.51) can be reduced to:

$$\frac{\partial \sigma_x}{\partial x} + \frac{\partial \tau_{yx}}{\partial y} + \bar{b}_x = 0 \tag{3.52}$$

Similarly, from the condition $\sum F_y = 0$, the following equation can be obtained:

$$\frac{\partial \sigma_y}{\partial y} + \frac{\partial \tau_{xy}}{\partial x} + \bar{b}_y = 0 \tag{3.53}$$

From the third condition, i.e., $\sum M_O = 0$, where O is the central point of the differential volume, the following equation can be obtained:

$$\begin{gathered}\left(\tau_{xy} + \frac{\partial \tau_{xy}}{\partial x}dx\right)dy \cdot t \cdot \frac{dx}{2} + \tau_{xy} dy \cdot t \cdot \frac{dx}{2} \\ - \left(\tau_{yx} + \frac{\partial \tau_{yx}}{\partial y}dy\right)dx \cdot t \cdot \frac{dy}{2} - \tau_{yx} dx \cdot t \cdot \frac{dy}{2} = 0\end{gathered} \tag{3.54}$$

The equation can be reduced to

$$\tau_{xy} = \tau_{yx} \tag{3.55}$$

In summary, the equilibrium equations for 2D case are

$$\begin{cases} \dfrac{\partial \sigma_x}{\partial x} + \dfrac{\partial \tau_{yx}}{\partial y} + \bar{b}_x = 0 \\ \dfrac{\partial \sigma_y}{\partial y} + \dfrac{\partial \tau_{xy}}{\partial x} + \bar{b}_y = 0 \\ \tau_{xy} = \tau_{yx} \end{cases} \tag{3.56}$$

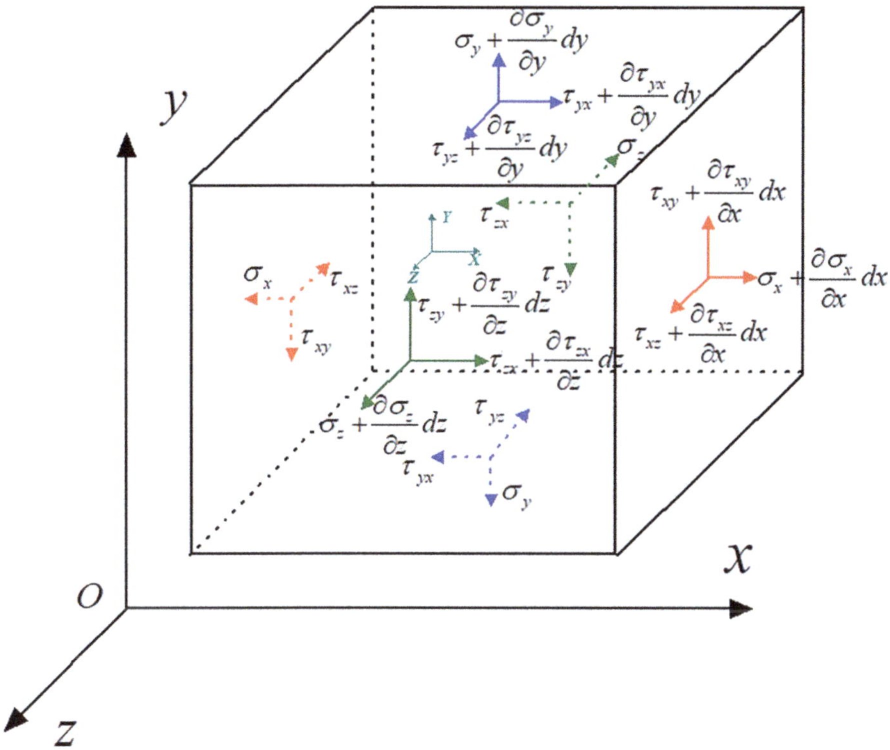

Fig. 3.18 The 3D differential volume

For the 3D case, the differential volume in Fig. 3.18 can be used to work out the equilibrium equations. The 3D differential volume is balanced and thus the following equilibrium conditions should be satisfied: (1) Summation of all the forces along the x, y and z directions is equal to zero; (2) Summation of all the moments along any point within the body is equal to zero.

From these conditions, the equilibrium equations for 3D case can be obtained:

$$\begin{cases} \dfrac{\partial \sigma_x}{\partial x} + \dfrac{\partial \tau_{yx}}{\partial y} + \dfrac{\partial \tau_{zx}}{\partial z} + \overline{b}_x = 0 \\ \dfrac{\partial \tau_{xy}}{\partial x} + \dfrac{\partial \sigma_y}{\partial y} + \dfrac{\partial \tau_{zy}}{\partial z} + \overline{b}_y = 0 \\ \dfrac{\partial \tau_{xz}}{\partial x} + \dfrac{\partial \tau_{yz}}{\partial y} + \dfrac{\partial \sigma_z}{\partial z} + \overline{b}_z = 0 \\ \tau_{xy} = \tau_{yx}, \tau_{xz} = \tau_{zx}, \tau_{zy} = \tau_{yz} \end{cases} \tag{3.57}$$

The first three equations can be rewritten in terms of index notation as:

$$\sigma_{ij,i} + \overline{b}_j = 0(i, j = x, y, z) \tag{3.58}$$

3.5 Calculation of the Principal Stress

We can again use the previous developments to discuss the issues of principal stresses and directions. It is known that the stress is a symmetric tensor. Using this fact, appropriate theory has been developed to identify and determine principal axes and values for the stress. For any given stress tensor, we can establish the principal value problem and solve the characteristic equation to explicitly determine the principal values and directions. The general characteristic equation for the stress tensor becomes

$$\det\left[\sigma_{ij} - \sigma\delta_{ij}\right] = -\sigma^3 + I_1\sigma^2 - I_2\sigma + I_3 = 0 \tag{3.59}$$

where $\sigma_1, \sigma_2, \sigma_3$ are the principal stresses and the fundamental invariants of the stress tensor can be expressed in terms of the three principal stresses $\sigma_1, \sigma_2, \sigma_3$ as

$$\begin{cases} I_1 = \sigma_1 + \sigma_2 + \sigma_3 \\ I_2 = \sigma_1\sigma_2 + \sigma_2\sigma_3 + \sigma_3\sigma_1 \\ I_3 = \sigma_1\sigma_2\sigma_3 \end{cases} \tag{3.60}$$

In the principal coordinate system, the stress matrix takes the special diagonal form

$$\sigma_{ij} = \begin{bmatrix} \sigma_1 & 0 & 0 \\ 0 & \sigma_2 & 0 \\ 0 & 0 & \sigma_3 \end{bmatrix} \tag{3.61}$$

A comparison of the general and principal stress states is shown in Fig. 3.19. Notice that for the principal coordinate system, all shearing stresses vanish and thus the state includes only normal stresses.

We now want to go back to investigate another issue related to stress and traction transformation that makes use of principal stresses. The general traction vector $\boldsymbol{T}^n$ that acts on an arbitrary surface as shown in Fig. 3.20 is considered. The issue of interest is to determine the traction vector's normal and shear components N and S. The normal component is simply the traction's projection in the direction of the unit normal vector $\boldsymbol{n}$, while the shear component is found by Pythagorean theorem:

$$\begin{cases} N = \boldsymbol{T}^n \cdot \boldsymbol{n} \\ S = \left(|\boldsymbol{T}^n|^2 - N^2\right)^{\frac{1}{2}} \end{cases} \tag{3.62}$$

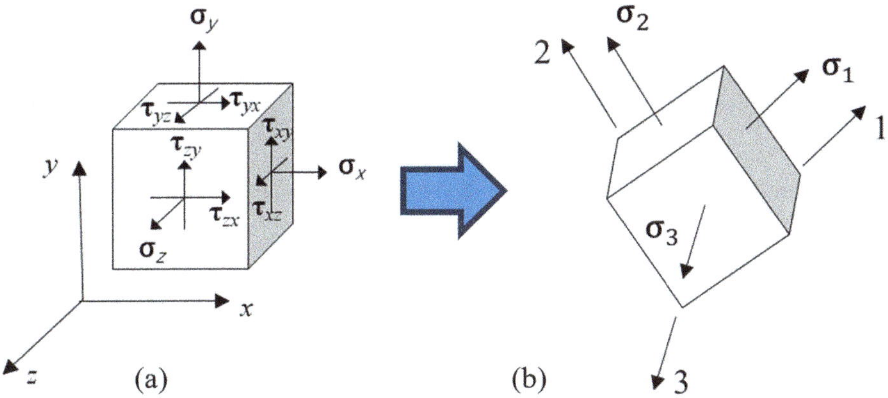

Fig. 3.19 Comparison of general and principal stress states: (**a**) General coordinate system; (**b**) Principal coordinate system

Fig. 3.20 Decomposition of traction vector into normal and shear directions

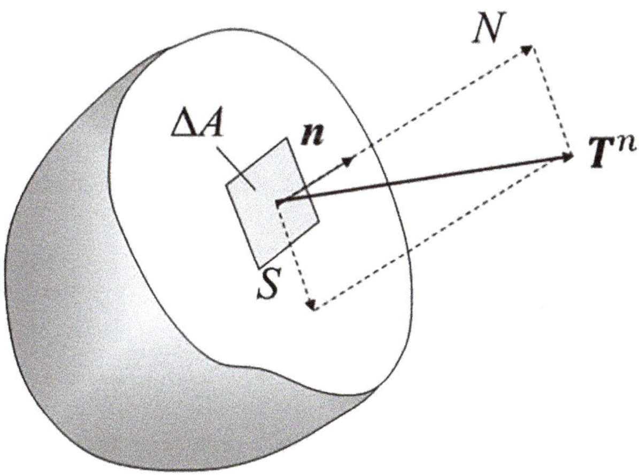

Putting the relationship for the traction vector Eq. (3.10) into Eq. (3.62) gives

$$\begin{aligned} N &= \boldsymbol{T^n} \cdot \boldsymbol{n} = T_i^n n_i = \sigma_{ji} n_j n_i \\ &= \sigma_1 n_1^2 + \sigma_2 n_2^2 + \sigma_3 n_3^2 \end{aligned} \tag{3.63}$$

where, in order to simplify the expressions, we have used the principal axes for the stress tensor. In a similar manner,

$$\begin{aligned} |\boldsymbol{T^n}|^2 &= \boldsymbol{T^n} \cdot \boldsymbol{T^n} = T_i^n T_i^n = \sigma_{ji} n_j \sigma_{ki} n_k \\ &= \sigma_1^2 n_1^2 + \sigma_2^2 n_2^2 + \sigma_3^2 n_3^2 \end{aligned} \tag{3.64}$$

Putting these results back into Eq. (3.62) yields

$$\begin{cases} N = \sigma_1 n_1^2 + \sigma_2 n_2^2 + \sigma_3 n_3^2 \\ S^2 + N^2 = \sigma_1^2 n_1^2 + \sigma_2^2 n_2^2 + \sigma_3^2 n_3^2 \end{cases} \tag{3.65}$$

In addition, we add the condition that the vector $\boldsymbol{n}$ has unit magnitude

$$1 = n_1^2 + n_2^2 + n_3^2$$

Equation (3.65) can be viewed as three linear algebraic equations for the unknowns n_1^2, n_2^2, n_3^2. Solving this system gives the following result:

$$\begin{cases} n_1^2 = \dfrac{S^2 + (N - \sigma_2)(N - \sigma_3)}{(\sigma_1 - \sigma_2)(\sigma_1 - \sigma_3)} \\ n_2^2 = \dfrac{S^2 + (N - \sigma_3)(N - \sigma_1)}{(\sigma_2 - \sigma_3)(\sigma_2 - \sigma_1)} \\ n_3^2 = \dfrac{S^2 + (N - \sigma_1)(N - \sigma_2)}{(\sigma_3 - \sigma_1)(\sigma_3 - \sigma_2)} \end{cases} \tag{3.66}$$

Without loss in generality, we can rank the principal stresses as $\sigma_1 > \sigma_2 > \sigma_3.$ Noting that the expressions given in Eq. (3.66) must be greater than or equal to zero, we can obtain the following relations

$$\begin{cases} S^2 + (N - \sigma_2)(N - \sigma_3) \geq 0 \\ S^2 + (N - \sigma_3)(N - \sigma_1) \leq 0 \\ S^2 + (N - \sigma_1)(N - \sigma_2) \geq 0 \end{cases} \tag{3.67}$$

For the equality case, Eq. (3.67) represents three circles in an S - N coordinate system, and Fig. 3.21 illustrates the location of each circle. These results are originally generated by Otto Mohr over a century ago, and the circles are commonly called *Mohr's circles of stress*. The three inequalities given in Eq. (3.67) imply that all admissible values of N and S lie in the shaded regions bounded by the three circles. Note that, for the ranked principal stresses, the largest shear component is easily determined as $S_{\max} = |\sigma_1 - \sigma_3|/2$. Although these circles can be effectively used for 2D stress transformation, the general tensorial-based relations given in Eq. (3.16) are normally used for general transformation computations.

There are several remarks related to the principal stresses. First, if the three principal stresses are not equal to each other, i.e., $\sigma_1 \neq \sigma_2 \neq \sigma_3$, then the three principal directions are orthogonal, which can be proved as follows. From Eq. (3.34), we have:

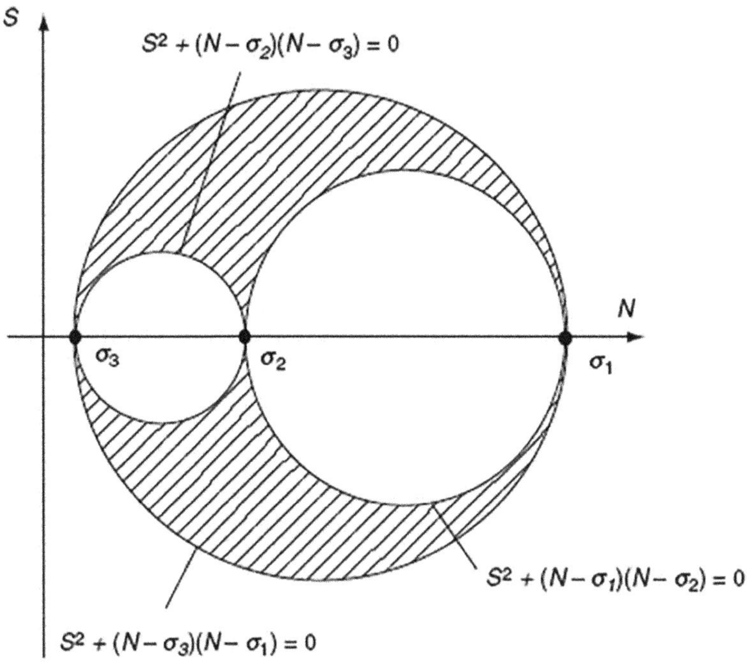

Fig. 3.21 Mohr's circle of stress

$$\begin{cases} (\sigma_x - \sigma_1)l_1 + \tau_{yx}m_1 + \tau_{zx}n_1 = 0 \\ \tau_{xy}l_1 + (\sigma_y - \sigma_1)m_1 + \tau_{zy}n_1 = 0 \\ \tau_{xz}l_1 + \tau_{yx}m_1 + (\sigma_z - \sigma_1)n_1 = 0 \\ (\sigma_x - \sigma_2)l_2 + \tau_{yx}m_2 + \tau_{zx}n_2 = 0 \\ \tau_{xy}l_2 + (\sigma_y - \sigma_2)m_2 + \tau_{zy}n_2 = 0 \\ \tau_{xz}l_2 + \tau_{yx}m_2 + (\sigma_z - \sigma_2)n_2 = 0 \\ (\sigma_x - \sigma_3)l_3 + \tau_{yx}m_3 + \tau_{zx}n_3 = 0 \\ \tau_{xy}l_3 + (\sigma_y - \sigma_3)m_3 + \tau_{zy}n_3 = 0 \\ \tau_{xz}l_3 + \tau_{yx}m_3 + (\sigma_z - \sigma_3)n_3 = 0 \end{cases} \tag{3.68}$$

From which we can obtain the following equations:

$$\begin{cases} (\sigma_1 - \sigma_2)(l_1 l_2 + m_1 m_2 + n_1 n_2) = 0 \\ (\sigma_1 - \sigma_3)(l_1 l_3 + m_1 m_3 + n_1 n_3) = 0 \\ (\sigma_2 - \sigma_3)(l_2 l_3 + m_2 m_3 + n_2 n_3) = 0 \end{cases} \tag{3.69}$$

Because $\sigma_1 \neq \sigma_2 \neq \sigma_3$, to make the equations above satisfied, the following equations should be satisfied:

$$\begin{cases} l_1 l_2 + m_1 m_2 + n_1 n_2 = 0 \\ l_1 l_3 + m_1 m_3 + n_1 n_3 = 0 \\ l_2 l_3 + m_2 m_3 + n_2 n_3 = 0 \end{cases} \tag{3.70}$$

which implying that the three principal directions, i.e., (l_1,m_1,n_1), (l_2,m_2,n_2), (l_3,m_3, n_3) are orthogonal.

The second remark is that if two of the three principal stresses are equal to each other, e.g., $\sigma_1 = \sigma_2 \neq \sigma_3$, then the principal direction of σ_3 is orthogonal to the directions of σ_1 and σ_2. The directions of σ_1 and σ_2 are not necessarily orthogonal, implying that any direction orthogonal to σ_3 may be the principal direction.

The third remark is that if the three principal stresses are equal to each other, i.e., $\sigma_1 = \sigma_2 = \sigma_3$, then the three principal directions are not necessarily orthogonal, implying that any direction may be the principal direction.

Example 3.1: Stress Transformation **Question:** For the following state of stress, determine the principal stresses and directions and find the traction vector on a plane with unit normal $\boldsymbol{n} = (0, 1, 1)/\sqrt{2}$.

$$\sigma_{ij} = \begin{bmatrix} 3 & 1 & 1 \\ 1 & 0 & 2 \\ 1 & 2 & 0 \end{bmatrix}$$

Solution: The principal stress problem is first solved by calculating the three invariants, giving the results $I_1 = 3$, $I_2 = -6$, $I_3 = -8$. This yields the following characteristic equation:

$$-\sigma^3 + 3\sigma^2 + 6\sigma - 8 = 0$$

The roots of this equation are: found to be $\sigma = 4, 1, -2$. Substituting the first root back into the fundamental system gives

$$\begin{cases} -n_1^{(1)} + n_2^{(1)} + n_3^{(1)} = 0 \\ n_1^{(1)} - 4n_2^{(1)} + 2n_3^{(1)} = 0 \\ n_1^{(1)} + 2n_2^{(1)} - 4n_3^{(1)} = 0 \end{cases}$$

Solving this system, the normalized principal direction is found to be $\boldsymbol{n}^{(1)} = (2, 1, 1)/\sqrt{6}$. In the similar fashion, the other two principal directions are $\boldsymbol{n}^{(2)} = (-1, 1, 1)/\sqrt{3}, \boldsymbol{n}^{(3)} = (0, -1, 1)/\sqrt{2}$. The traction vector on the specified plane is calculated as below

$$T_i^n = \begin{bmatrix} 3 & 1 & 1 \\ 1 & 0 & 2 \\ 1 & 2 & 0 \end{bmatrix} \begin{bmatrix} 0 \\ 1/\sqrt{2} \\ 1/\sqrt{2} \end{bmatrix} = \begin{bmatrix} 2/\sqrt{2} \\ 2/\sqrt{2} \\ 2/\sqrt{2} \end{bmatrix}$$

3.6 Calculation of Maximal Shear Stress

In the previous part, the calculation process for the principal stress is presented. One of the reasons that the principal stresses are calculated is to be used as the failure criterion of materials. However, for some materials, the principal stresses are not the appropriate parameters for serving as the failure criterion. Instead, the shear stress may be more suitable for determining whether failure has occurred in the material or not. Therefore, in this part, the calculation process for the maximal shear stress is presented.

The differential tetrahedron in Fig. 3.22 is used to illustrate the calculation of the maximal shear stress. In the three orthogonal differential planes, i.e., *Oab*, *Oac* and *Obc*, there are only principal stresses. The task is to find the maximal shear stress at the oblique plane *abc*. The normal direction of the oblique plane is represented as v and the three stress components are f_{vx}, f_{vy} and f_{vz}. From Eq. (3.34), the relation between f_{vx}, f_{vy}, f_{vz} and the three principal stresses σ_1, σ_2, σ_3 are:

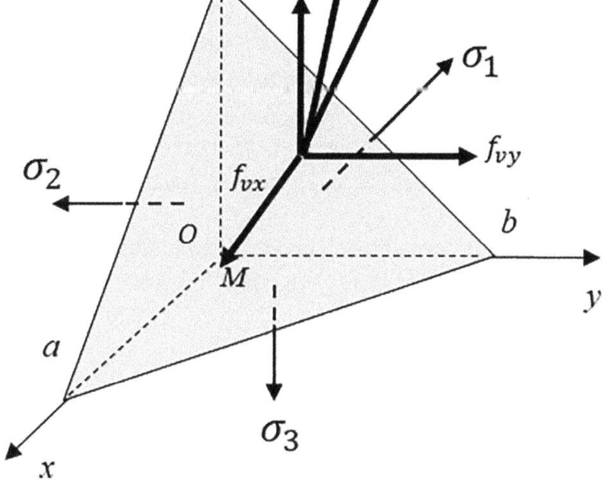

Fig. 3.22 Calculation of the maximal shear stress

$$f_{vx}=\sigma_1 l, f_{vy}=\sigma_2 m, f_{vz}=\sigma_3 n \tag{3.71}$$

where, l, m and n are the cosines from $\boldsymbol{v}$ to the x, y and z axes, i.e.,

$$\cos(\boldsymbol{v},x)=l,\ \cos(\boldsymbol{v},y)=m,\ \cos(\boldsymbol{v},z)=n \tag{3.72}$$

the traction at the oblique plane can be worked out as:

$$f_v=\sqrt{f_{vx}^2+f_{vy}^2+f_{vz}^2}=\sqrt{\sigma_1^2 l^2+\sigma_2^2 m^2+\sigma_3^2 n^2} \tag{3.73}$$

The traction vector at the oblique plane can be projected to the normal and shear directions. The normal stress at the oblique plane σ_v can be worked out as:

$$\sigma_v=f_{vx}l+f_{vy}m+f_{vz}n=\sigma_1 l^2+\sigma_2 m^2+\sigma_3 n^2 \tag{3.74}$$

Then the shear stress at the oblique plane τ_v can be worked out from the traction vector and the normal stress as:

$$\tau_v^2=f_v^2-\sigma_v^2=\sigma_1^2 l^2+\sigma_2^2 m^2+\sigma_3^2 n^2-\left(\sigma_1 l^2+\sigma_2 m^2+\sigma_3 n^2\right)^2 \tag{3.75}$$

The expression for the shear stress at the oblique plane, i.e., Eq. (3.75), has been established. Now the task is to find the maximal value of Eq. (3.75). Because $l^2+m^2+n^2=1$, i.e., only two of l, m and n are independent, one variable can be represented using the other two variables. Here, we eliminate the variable n, then we have:

$$\tau_v^2=\left(\sigma_1^2-\sigma_3^2\right)l^2+\left(\sigma_2^2-\sigma_3^2\right)m^2+\sigma_3^2-\left[(\sigma_1-\sigma_3)l^2+(\sigma_2-\sigma_3)m^2+\sigma_3^2\right] \tag{3.76}$$

To find the maximal value of the expression above, the partial derivatives of Eq. (3.76) with respective to the variables l and m, i.e., $\frac{\partial}{\partial l}\left(\tau_v^2\right)=0$, $\frac{\partial}{\partial m}\left(\tau_v^2\right)=0$, from which we can obtain the following two equations:

$$\begin{cases}\left(\sigma_1^2-\sigma_3^2\right)l-2\left[(\sigma_1-\sigma_3)l^2+(\sigma_2-\sigma_3)m^2+\sigma_3^2\right](\sigma_1-\sigma_3)l=0\\ \left(\sigma_2^2-\sigma_3^2\right)m-2\left[(\sigma_1-\sigma_3)l^2+(\sigma_2-\sigma_3)m^2+\sigma_3^2\right](\sigma_2-\sigma_3)m=0\end{cases} \tag{3.77}$$

The two equations above can be solved under different cases. In the first case, all the three principal stresses are not equal to each other, i.e., $\sigma_1\neq\sigma_2\neq\sigma_3$, then Eq. (3.77) can be rewritten as:

$$\left.\begin{aligned}\left\{(\sigma_1-\sigma_3)-2\left[(\sigma_1-\sigma_3)l^2+(\sigma_2-\sigma_3)m^2\right]\right\}l=0\\ \left\{(\sigma_2-\sigma_3)-2\left[(\sigma_1-\sigma_3)l^2+(\sigma_2-\sigma_3)m^2\right]\right\}m=0\end{aligned}\right\} \tag{3.78}$$

Table 3.6 Solutions of the maximal shear stress

l	0	0	± 1	0	$\pm \frac{1}{\sqrt{2}}$	$\pm \frac{1}{\sqrt{2}}$
m	0	± 1	0	$\pm \frac{1}{\sqrt{2}}$	0	$\pm \frac{1}{\sqrt{2}}$
n	± 1	0	0	$\pm \frac{1}{\sqrt{2}}$	$\pm \frac{1}{\sqrt{2}}$	0
τ_v	0	0	0	$\pm \frac{\sigma_2 - \sigma_3}{2}$	$\pm \frac{\sigma_3 - \sigma_1}{2}$	$\pm \frac{\sigma_1 - \sigma_2}{2}$

One solution of the equations above is $l = 0$, $m = 0$, $n = \pm 1$, which gives $\tau_v^2 = 0$, implying there is no maximal shear stress in this scenario. The second solution is $l = 0, m = \pm \frac{1}{\sqrt{2}}, \ n = \pm \frac{1}{\sqrt{2}},$ which gives $\tau_v = \pm \frac{\sigma_2 - \sigma_3}{2}$. The third solution is $l = \pm \frac{1}{\sqrt{2}}, m = 0, \ n = \pm \frac{1}{\sqrt{2}},$ which gives $\tau_v = \pm \frac{\sigma_3 - \sigma_1}{2}$.

In the second case, two of the principal stresses are equal to each other. Here we set $\sigma_1 = \sigma_3 > \sigma_2$, then Eq. (3.77) can be rewritten as:

$$\left.\begin{array}{l} \{(\sigma_1 - \sigma_3) - 2[(\sigma_1 - \sigma_3)l^2 + (\sigma_2 - \sigma_3)m^2]\}l = 0 \\ \{(\sigma_2 - \sigma_3) - 2[(\sigma_1 - \sigma_3)l^2 + (\sigma_2 - \sigma_3)m^2]\}m = 0 \end{array}\right\} \tag{3.79}$$

The second equation above can be reduced to:

$$\{(\sigma_2 - \sigma_3) - 2[(\sigma_2 - \sigma_3)m^2]\}m = 0 \tag{3.80}$$

The first solution of the equation above is $m = 0$, which gives $\tau_v = 0$, implying there is no maximal shear stress in this scenario. The second solution is $m = \pm \frac{1}{\sqrt{2}},$ which gives $\tau_v = \pm \frac{\sigma_1 - \sigma_2}{2}$.

The third case is that all the three principal stresses are equal to each other, i.e., $\sigma_1 = \sigma_2 = \sigma_3$, then from Eq. (3.76), we can have $\tau_v = 0$, implying there is no maximal shear stress in this scenario.

In summary, the solutions of the maximal shear stress for all the cases are presented in Table 3.6.

3.7 Some Stress Definitions

As mentioned in our previous discussion on strain, it is often convenient to decompose the stress into two parts called the *spherical and deviatoric stress tensors*. Analogous to Eqs. (2.36) and (2.37), the *spherical stress* is defined by

$$\widetilde{\sigma}_{ij} = \frac{1}{3}\sigma_{kk}\delta_{ij} \tag{3.81}$$

while the *deviatoric stress* becomes

$$\hat{\sigma}_{ij} = \sigma_{ij} - \frac{1}{3}\sigma_{\overline{kk}}\delta_{ij} \tag{3.82}$$

Note that the total stress is then simply the sum of the *spherical and deviatoric stresses*, i.e.

$$\sigma_{ij} = \overline{\sigma}_{ij} + \hat{\sigma}_{ij} \tag{3.83}$$

The spherical stress is an isotropic tensor, being the same in all coordinate systems. It can be shown that the principal directions of the deviatoric stress are the same as those of the stress tensor itself. We next briefly explore a couple of particular stress components or combinations that have been defined in the literature and are commonly used in formulating failure theories related to inelastic deformation. It should be noted that ductile materials normally exhibit inelastic yielding failures that can be characterized by these particular stresses.

First the normal and shear stresses (tractions) that act on a special plane the normal of which makes equal angles with the three principal axes are considered. This plane is commonly referred to as the octahedral plane. Determination of these normal and shear stresses is straightforward if we use the principal axes of stress. Since the unit normal vector to the octahedral plane makes equal angles with the principal axes, its components are given by $n_i = \pm(1,1,1)/\sqrt{3}$. Referring to Fig. 3.22 and using the results of the previous section give the desired normal and shear stresses as

$$\begin{cases} N = \sigma_{oct} = \frac{1}{3}(\sigma_1 + \sigma_2 + \sigma_3) = \frac{1}{3}\sigma_{kk} = \frac{1}{3}I_1 \\ S = \tau_{oct} = \frac{1}{3}\left[(\sigma_1 - \sigma_2)^2 + (\sigma_2 - \sigma_3)^2 + (\sigma_3 - \sigma_1)^2\right]^{\frac{1}{2}} \\ \quad = \frac{1}{3}\left(2I_1^2 - 6I_2\right)^{\frac{1}{2}} \end{cases} \tag{3.84}$$

It can be shown that the octahedral shear stress τ_{oct} is directly related to the distortional strain energy, which is often used in failure theories for ductile materials.

Another specially defined stress also related to the distortional strain energy failure criteria is known as the effective or *von Mises stress* and is given by the expression

$$\sigma_e = \sigma_{\text{von Mises}} = \frac{1}{\sqrt{2}}\left[(\sigma_1 - \sigma_2)^2 + (\sigma_2 - \sigma_3)^2 + (\sigma_3 - \sigma_1)^2\right]^{\frac{1}{2}} \tag{3.85}$$

Note that although the von Mises stress is not really a particular stress or traction component in the usual sense, it is obviously directly related to the octahedral shear stress by the relation $\boldsymbol{\sigma_e} = \left(\mathbf{3}/\sqrt{\mathbf{2}}\right)\boldsymbol{\tau_{oct}}$. If at some point in the structure, the von Mises stress equals the yield stress, then the material is considered to be at the failure condition. Based on this fact, many finite element computer codes commonly plot von Mises stress distributions based on the numerically generated stress field. It should be noted that the von Mises and octahedral shear stresses involve only the differences in the principal stresses and not the individual values. Thus, increasing each principal stress by the same amount will not change the value of $\boldsymbol{\sigma_e}$ or $\boldsymbol{\tau_{oct}}$. This result also implies that these values are independent of the hydrostatic stress.

Chapter 4
Constitutive Relation

Yongtao Lyu and Yonggang Zheng

4.1 Introduction

The previous two chapters established the elasticity field equations related to the kinematics of small deformation theory and the equilibrium of the associated internal stress field. Based on these physical concepts, six strain-displacement relations, six compatibility equations and three equilibrium equations can be developed for the general 3D case. Because moment equilibrium simply results in symmetry of the stress tensor, it is not normally included as a separate field equation set. Also, recall that the compatibility equations actually represent only three independent relations, and these equations are needed only to ensure that a given strain field will produce single-valued continuous displacements. Because the displacements are included in the general problem formulation, the solution normally gives continuous displacements, and the compatibility equations are not formally needed for the general system. Thus, excluding the compatibility relations, nine field equations have been developed. The unknowns in these equations include three displacement components, six components of strain, and six stress components, yielding a total of fifteen unknowns. Thus, the nine equations are not sufficient to work out the fifteen unknowns, and additional field equations are needed. This result should not be surprising sincel up to this point in our development we have not considered the material response.

We now want to complete our general formulations by specializing to a particular material model that provides reasonable characterization of materials under small deformations. The model we will use is a linear elastic material. This chapter

Y. Lyu (✉) · Y. Zheng
School of Mechanics and Aerospace Engineering, Dalian University of Technology, Dalian, Liaoning, China
e-mail: Yongtaolu@dlut.edu.cn; zhengyg@dlut.edu.cn

Y. Lyu, Y. Zheng, *Theory of Elasticity*,
https://doi.org/10.1007/978-981-95-6043-1_4

presents the basics of the elastic model specializing the formulation for isotropic elastic body.

4.2 Material Models

Relations that characterize the physical properties of materials are called *constitutive equations*. Because of the endless variety of materials and loadings, the study and development of constitutive equations is perhaps one of the most interesting and challenging fields in mechanics. Although the theory of continuum mechanics has established some principles for systematic development of constitutive equations, many constitutive laws have been developed through empirical relations based on experimental evidence. Our interest here is limited to a special class of solid materials with loadings resulting from mechanical or thermal effects. The mechanical behavior of solids is normally defined by constitutive (stress-strain) relations. Commonly, these relations express the stress as a function of the strain, strain rate, strain history, temperature and material properties. We choose a rather simple material model called the elastic solid that does not include rate or history effects. The model may be described as a deformable continuum that recovers its original configuration when the loadings causing the deformation are removed. Furthermore, we restrict the constitutive stress-strain law to be linear, thus leading to a linear elastic solid. Although these assumptions greatly simplify the model, linear elasticity predictions have shown a good agreement with experimental data and have provided useful methods to conduct stress analysis. Many structural materials including metals, plastics, ceramics, wood, rock, concrete, and so forth exhibit linear elastic behaviors under small deformations.

As mentioned, experimental testing is commonly employed to characterize the mechanical behavior of materials. One such technique is the simple tension test in which a specially prepared cylindrical or flat stock sample is loaded axially in a testing machine. Strain is determined by the change in length between prescribed reference marks on the sample and is usually measured by a clip gauge. The loading data collected from a load cell is divided by the cross-sectional area in the test section to calculate the stress. Axial stress-strain data is recorded and plotted using standard experimental techniques. Typical qualitative data for three types of structural metals (mild steel, aluminum, cast iron) are shown in Fig. 4.1. It is observed that each material exhibits an initial stress-strain response for small deformation that is approximately linear. This is followed by a change to nonlinear behavior that can lead to large deformation, finally ending with failure of the sample.

For each material, the initial linear response ends at a point normally referred to as the proportional limit. Another observation in this initial region is that if the loading is removed, the sample returns to its original shape and the strain disappears. This characteristic is the primary descriptor of elastic behavior. However, at some point on the stress-strain curve, unloading does not bring the sample back to zero strain and some permanent plastic deformation occurs. The point at which this non-elastic

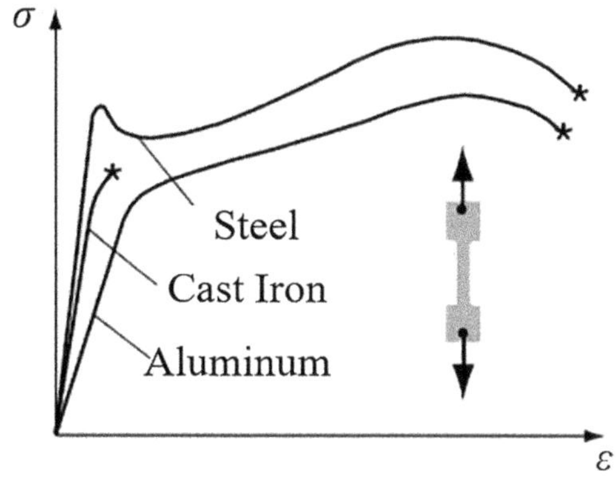

Fig. 4.1 Typical uniaxial stress-strain curves for three structural metals

behavior begins is called the *elastic limit*. Although some materials exhibit different elastic and proportional limits, many times these values are taken to be approximately the same. Another demarcation on the stress-strain curve is referred to as the yield point, defined by the location where large plastic deformation begins.

Because mild steel and aluminum are ductile materials, their stress-strain response indicates extensive plastic deformation, and during this period, dimensions of the sample will change. In particular, the sample's cross-sectional area undergoes significant reduction, and the calculation of stress using the division by the original area will now be in error. This accounts for the reduction in the stress at large strain. If we were to calculate the load divided by the true area, the true stress would continue to increase until failure. On the other hand, cast iron is known to be a brittle material, and thus its stress-strain response does not show large plastic deformation. For this material, very little non-elastic or nonlinear behavior is observed. It is therefore concluded from this and many other studies that a large variety of materials exhibit linear elastic behavior under small deformations. This would lead to a linear constitutive model for the 1D axial loading case given by the relation $\sigma = E\varepsilon$, where E is the slope of the uniaxial stress-strain curve. We now use this simple concept to develop the general 3D forms of the linear elastic constitutive model.

4.3 Generalized Hooke's Law

Based on observations from the previous section, to construct a general 3D constitutive law for linear elastic materials, we assume that each stress component is linearly related to each strain component

$$\begin{cases} \sigma_x = C_{11}\varepsilon_x + C_{12}\varepsilon_y + C_{13}\varepsilon_z + 2C_{14}\varepsilon_{xy} + 2C_{15}\varepsilon_{yz} + 2C_{16}\varepsilon_{zx} \\ \sigma_y = C_{21}\varepsilon_x + C_{22}\varepsilon_y + C_{23}\varepsilon_z + 2C_{24}\varepsilon_{xy} + 2C_{25}\varepsilon_{yz} + 2C_{26}\varepsilon_{zx} \\ \sigma_z = C_{31}\varepsilon_x + C_{32}\varepsilon_y + C_{33}\varepsilon_z + 2C_{34}\varepsilon_{xy} + 2C_{35}\varepsilon_{yz} + 2C_{36}\varepsilon_{zx} \\ \tau_{xy} = C_{41}\varepsilon_x + C_{42}\varepsilon_y + C_{43}\varepsilon_z + 2C_{44}\varepsilon_{xy} + 2C_{45}\varepsilon_{yz} + 2C_{46}\varepsilon_{zx} \\ \tau_{yz} = C_{51}\varepsilon_x + C_{52}\varepsilon_y + C_{53}\varepsilon_z + 2C_{54}\varepsilon_{xy} + 2C_{55}\varepsilon_{yz} + 2C_{56}\varepsilon_{zx} \\ \tau_{zx} = C_{61}\varepsilon_x + C_{62}\varepsilon_y + C_{63}\varepsilon_z + 2C_{64}\varepsilon_{xy} + 2C_{65}\varepsilon_{yz} + 2C_{66}\varepsilon_{zx} \end{cases} \tag{4.1}$$

where the coefficients C_{ij} are material parameters and the factors of 2 arise because of the symmetry of the strain. Note that this relation could also be expressed by expressing the strains as a linear function of the stress components. These relations can be described into a matrix format as

$$\begin{bmatrix} \sigma_x \\ \sigma_y \\ \sigma_z \\ \tau_{xy} \\ \tau_{yz} \\ \tau_{zx} \end{bmatrix} = \begin{bmatrix} C_{11} & C_{12} & \cdot & \cdot & \cdot & C_{16} \\ C_{21} & \cdot & \cdot & \cdot & \cdot & \cdot \\ \cdot & \cdot & \cdot & \cdot & \cdot & \cdot \\ \cdot & \cdot & \cdot & \cdot & \cdot & \cdot \\ \cdot & \cdot & \cdot & \cdot & \cdot & \cdot \\ C_{61} & \cdot & \cdot & \cdot & \cdot & C_{66} \end{bmatrix} \begin{bmatrix} \varepsilon_x \\ \varepsilon_y \\ \varepsilon_z \\ 2\varepsilon_{xy} \\ 2\varepsilon_{yz} \\ 2\varepsilon_{zx} \end{bmatrix} \tag{4.2}$$

The relation above can also be expressed in standard tensor notation by writing

$$\sigma_{ij} = C_{ijkl}\varepsilon_{kl} \tag{4.3}$$

where C_{ijkl} is a fourth-order elasticity tensor whose components include all the material parameters necessary to characterize the material. Based on the symmetry of the stress and strain tensors, the elasticity tensor must have the following properties:

$$\begin{cases} C_{ijkl} = C_{jikl} \\ C_{ijkl} = C_{ijlk} \end{cases} \tag{4.4}$$

In general, the fourth-order tensor C_{ijkl} has 81 components. However, Eq. (4.1) reduces the number of independent components to 36. Later in this textbook we introduce the concept of strain energy, and this leads to the relation $C_{ijkl} = C_{klij}$ or equivalently $C_{ij} = C_{ji}$, which provides further reduction to 21 independent elastic components. The components of C_{ijkl} or equivalently C_{ij} are called *elastic stiffness components* and have units of stress (force/area). In order to continue further, we must address the issues of material homogeneity and isotropy.

If the material is homogeneous, the elastic behavior does not vary spatially, and thus all elastic moduli are constant. For this case, the elasticity formulation is straightforward, leading to the development of many analytical solutions to problems of engineering interest. A homogeneous assumption is an appropriate model for most structural applications, and thus we primarily choose this particular case for subsequent formulation and problem solution.

Studies in geo-mechanics have found that the material behavior of soil and rock commonly depends on the distance below the earth's surface. In order to simulate particular geo-mechanical problems, researchers have used nonhomogeneous elastic models applied to semi-infinite domains. Typical applications have involved modeling the response of a semi-infinite soil mass under surface or subsurface loadings with variation in elastic modulus. Another more recent application involves the behavior of functionally graded materials (FGM). FGMs are a new class of engineered materials developed with spatially varying properties to suit particular applications. The graded composition of such materials is commonly established and controlled using powder metallurgy, chemical vapor deposition, or centrifugal casting. Typical analytical studies of these materials have assumed linear, exponential, and power-law variation in elastic modulus of the form as below:

$$\begin{cases} C_{ij}(x) = C_{ij}^{0}(1 + \mathrm{a}x) \\ C_{ij}(x) = C_{ij}^{0}e^{\mathrm{a}x} \\ C_{ij}(x) = C_{ij}^{0}x^{\mathrm{a}} \end{cases} \tag{4.5}$$

where C_{ij}^{0} and a are the prescribed constants and x is the spatial coordinate.

Similar to homogeneity, another fundamental material property is isotropy. This property has to do with differences in material modulus with respect to orientation. For example, many materials, including crystalline minerals, wood and fiber-reinforced composites, have different elastic moduli in different directions. Such materials are said to be anisotropic. Note that for most anisotropic materials, there exist particular directions where the properties are the same. These directions indicate material symmetries. However, for many engineering materials (most structural metals and many plastics), the orientation of crystalline and grain microstructure is distributed randomly so that macroscopic elastic properties are found to be essentially the same in all directions. Such materials with complete symmetry are called *isotropic material*. As expected, an anisotropic model complicates the formulation and solution of problems. We therefore postpone the development of such solutions and continue our current development under the assumption of isotropic material behavior.

The tensor form provides a convenient way to establish the desired isotropic stress-strain relations. If we assume isotropic behavior, the elasticity tensor must be the same under all rotations of the coordinate system. Using the basic transformation properties, the fourth-order elasticity tensor must satisfy

$$C_{ijkl} = Q_{im}Q_{jn}Q_{kp}Q_{lq}C_{mnpq} \tag{4.6}$$

It can be seen that the most general form that satisfies this isotropy condition can be given by

$$C_{ijkl} = \alpha\delta_{ij}\delta_{kl} + \beta\delta_{ik}\delta_{jl} + \gamma\delta_{il}\delta_{jk} \tag{4.7}$$

where α, β, and γ are arbitrary constants. Using the general form in the stress-strain relation gives

$$\sigma_{ij} = \lambda\varepsilon_{kk}\delta_{ij} + 2\mu\varepsilon_{ij} \tag{4.8}$$

where we have relabeled particular constants using λ and μ. The elastic constant λ is called *Lame's constant*, and μ is referred to as the shear modulus or modulus of rigidity. Some textbooks use the notation G for the shear modulus. Equation (4.8) can be expressed in individual scalar format as

$$\begin{cases} \sigma_x = \lambda(\varepsilon_x + \varepsilon_y + \varepsilon_z) + 2\mu\varepsilon_x \\ \sigma_y = \lambda(\varepsilon_x + \varepsilon_y + \varepsilon_z) + 2\mu\varepsilon_y \\ \sigma_z = \lambda(\varepsilon_x + \varepsilon_y + \varepsilon_z) + 2\mu\varepsilon_z \\ \tau_{xy} = 2\mu\varepsilon_{xy} \\ \tau_{yz} = 2\mu\varepsilon_{yz} \\ \tau_{zx} = 2\mu\varepsilon_{zx} \end{cases} \tag{4.9}$$

Relations in Eq. (4.9) are called the *generalized Hooke's law* for linear isotropic elastic solids. They are named after Robert Hooke who in 1678 first proposed that the deformation of an elastic structure is proportional to the applied force. It should be noted that only two independent elastic constants are needed to describe the mechanical behavior of isotropic materials. It should be noted that additional numbers of elastic moduli are needed in the corresponding relations for anisotropic materials.

Stress-strain relations given in Eq. (4.8) may be inverted to express the strain in terms of the stress. In order to do this, it is convenient to use the index notation format and set the two free indices the same (contraction process) to get

$$\sigma_{kk} = (3\lambda + 2\mu)\varepsilon_{kk} \tag{4.10}$$

This relation can be solved for ε_{kk} and substituted back into Eq. (4.8) to get

$$\varepsilon_{ij} = \frac{1}{2\mu}\left(\sigma_{ij} - \frac{\lambda}{3\lambda + 2\mu}\sigma_{kk}\delta_{ij}\right) \tag{4.11}$$

which is more commonly written as

$$\varepsilon_{ij} = \frac{1+\nu}{E}\sigma_{ij} - \frac{\nu}{E}\sigma_{kk}\delta_{ij} \tag{4.12}$$

where $E = \mu(3\lambda + 2\mu)/(\lambda + \mu)$ and is called the *modulus of elasticity* or *Young's modulus*, and $\nu = \lambda/[2(\lambda + \mu)]$ is referred to as *Poisson's ratio*. The relation given in the index notation (Eq. (4.12)) may be written out in component (scalar) format, giving the six equations

$$\begin{cases} \varepsilon_x = \dfrac{1}{E}\left[\sigma_x - \nu\left(\sigma_y + \sigma_z\right)\right] \\ \varepsilon_y = \dfrac{1}{E}\left[\sigma_y - \nu\left(\sigma_z + \sigma_x\right)\right] \\ \varepsilon_z = \dfrac{1}{E}\left[\sigma_z - \nu\left(\sigma_x + \sigma_y\right)\right] \\ \varepsilon_{xy} = \dfrac{1+\nu}{E}\tau_{xy} = \dfrac{1}{2\mu}\tau_{xy} \\ \varepsilon_{yz} = \dfrac{1+\nu}{E}\tau_{yz} = \dfrac{1}{2\mu}\tau_{yz} \\ \varepsilon_{zx} = \dfrac{1+\nu}{E}\tau_{zx} = \dfrac{1}{2\mu}\tau_{zx} \end{cases} \tag{4.13}$$

Constitutive relation given in Eq. (4.13) again illustrates that only two elastic constants are needed to formulate Hooke's law for isotropic materials. By using any of the isotropic forms of Hooke's law, it can be shown that the principal axes of stress coincide with the principal axes of strain. This result also holds for some but not all anisotropic materials.

4.4 Constitutive Relation for Elastic Solids

In this part, the constitutive relations for different elastic solids are explored, starting from the most complex one (fully anisotropic elastic solid) to the simplest one (isotropic elastic solid).

It has long been recognized that deformation behavior of many materials depends upon the orientation; that is, the stress-strain response of a sample taken from the material in one direction will be different if the sample are taken in a different direction. The term anisotropy is generally used to describe such behaviors. Early investigators of these phenomena are motivated by the response of naturally occurring anisotropic materials such as wood and crystalline solids. Today, extensive use of engineered composites has brought forward many new types of fiber- and particle-reinforced materials with anisotropic response. Thus, knowledge of stress distributions in anisotropic materials is very important for proper use of these new high-performance materials in structural applications. Our previous development of the linear elastic stress-strain relations began with the general case of inhomogeneous and anisotropic behavior. However, this generality was quickly eliminated and only the homogeneous isotropic case was subsequently developed in detail. We now want to go back and further investigate the anisotropic homogeneous case and develop applications for particular elasticity problems including torsion and plane problems.

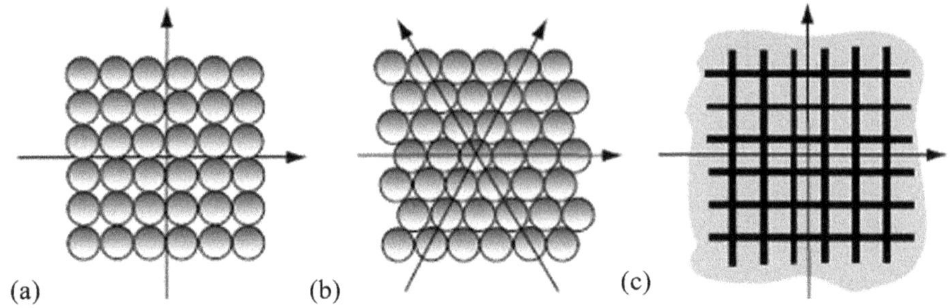

Fig. 4.2 Microstructures of different materials: (**a**) Simple cubic crystal; (**b**) Hexagonal crystal; (**c**) Fiber reinforced composite (arrows indicate material symmetry directions)

The directional-dependent behaviors found in anisotropic solids normally result from particular microstructural features within the material. Our previous isotropic model neglected these effects, thus resulting in a material that behaved in the same way in all directions. Micro features commonly arise in natural and synthetic materials in such a way as to produce a stress-strain response with particular symmetries. This concept is based on the Neumann principle that symmetry in material microgeometry corresponds to identical symmetry in the constitutive response. We can qualitatively gain an understanding of this concept from some simple 2D cases shown in Fig. 4.2. The figure illustrates idealized internal microstructures of two crystalline solids and one fiber reinforced composite. The two crystalline materials correspond to special atomic packing arrangements that lead to identical behaviors in the indicated directions of the arrows. The fiber-reinforced composite has a 90° fiber layout, which again produces identical behaviors in the layout directions. Many other deformation symmetries exist for more complicated microstructures and some follow a curvilinear reference system such as that found in wood. These symmetries generally lead to a reduction in the complexity of the stress-strain constitutive relation. On a relevant topic, for multiphase or porous materials, many researchers have been trying to establish relationships between the elasticity tensor C_{ijkl} and a fabric tensor that characterizes the arrangement of microstructural material components.

The general form of Hooke's law was given by

$$\sigma_{ij} = C_{ijkl}\varepsilon_{kl} \tag{4.14}$$

The fourth-order elasticity tensor C_{ijkl} contains all of the elastic stiffness components, and we have previously established the following symmetric properties:

$$C_{ijkl} = C_{jikl} = C_{ijlk} = C_{klij} \tag{4.15}$$

The first two symmetries in relation above come from the symmetry of the stress and strain tensors, while the last relation comes from arguments based on the existence of the strain energy function. Eq. (4.15) reduces the original 81 independent elastic constants within C_{ijkl} to a set of 21 elastic constants for the general case. We shall assume that the material is homogeneous and thus the elastic moduli are independent of the spatial position. On occasion we may want to invert Eq. (4.14) and write out the strains using the stresses, i.e.

$$\varepsilon_{ij} = S_{ijkl}\sigma_{kl} \tag{4.16}$$

where S_{ijll} is the elastic compliance tensor, which has symmetric properties same as those in Eq. (4.14).

Because of the various pre-existing symmetries, stress-strain relations presented in Eq. (4.14) contain many superfluous terms and equations. To avoid these, a convenient contracted notation has been developed, sometimes referred to as Voigt matrix notation.

$$\begin{bmatrix} \sigma_x \\ \sigma_y \\ \sigma_z \\ \tau_{yz} \\ \tau_{zx} \\ \tau_{xy} \end{bmatrix} = \begin{bmatrix} C_{11} & C_{12} & \cdots & C_{16} \\ C_{21} & \cdot & \cdots & \cdot \\ \cdot & \cdot & \cdots & \cdot \\ \cdot & \cdot & \cdots & \cdot \\ \cdot & \cdot & \cdots & \cdot \\ C_{61} & \cdot & \cdots & C_{66} \end{bmatrix} \begin{bmatrix} \varepsilon_x \\ \varepsilon_y \\ \varepsilon_z \\ 2\varepsilon_{yz} \\ 2\varepsilon_{zx} \\ 2\varepsilon_{xy} \end{bmatrix} \tag{4.17}$$

or in compact notation

$$\sigma_i = C_{ij}\varepsilon_j \tag{4.18}$$

where σ_i and ε_i are the first-order tensors. Note that the symmetry imposed by strain energy implies that the matrix $\boldsymbol{C}$ is symmetric; that is, $C_{ij} = C_{ji}$, and thus only 21 independent elastic stiffness components exist. The σ_i and ε_i are related by the following symmetric matrix:

$$C_{ij} = \begin{bmatrix} C_{1111} & C_{1122} & C_{1133} & C_{1123} & C_{1131} & C_{1112} \\ \cdot & C_{2222} & C_{2233} & C_{2223} & C_{2231} & C_{2212} \\ \cdot & \cdot & C_{3333} & C_{3323} & C_{3331} & C_{3312} \\ \cdot & \cdot & \cdot & C_{2323} & C_{2331} & C_{2312} \\ \cdot & Sym & \cdot & \cdot & C_{3131} & C_{3112} \\ \cdot & \cdot & \cdot & \cdot & \cdot & C_{1212} \end{bmatrix} \tag{4.19}$$

A similar scheme can be established for Eq. (4.16), and an elastic compliance matrix S_{ij} can be defined as

$$\varepsilon_i = S_{ij}\sigma_j \tag{4.20}$$

where,

$$S_{ij} = \begin{bmatrix} S_{1111} & S_{1122} & S_{1133} & S_{1123} & S_{1131} & S_{1112} \\ \cdot & S_{2222} & S_{2233} & S_{2223} & S_{2231} & S_{2212} \\ \cdot & \cdot & S_{3333} & S_{3323} & S_{3331} & S_{3312} \\ \cdot & \cdot & \cdot & S_{2323} & S_{2331} & S_{2312} \\ \cdot & Sym & \cdot & \cdot & S_{3131} & S_{3112} \\ \cdot & \cdot & \cdot & \cdot & \cdot & S_{1212} \end{bmatrix}$$

4.4.1 Material Symmetry

From the previous section, we determined that for the general anisotropic case (sometimes referred to as triclinic material), 21 independent elastic constants are needed to characterize the material response. However, most engineering materials have some types of symmetry, which further reduces the required number of independent elastic constants. Orientations for which an anisotropic material has the same stress-strain response can be determined by coordinate transformation (rotation) theory previously discussed. Such particular transformations are sometimes called the material symmetry group. In order to determine various material symmetries, it is more convenient to work in the non-contracted form. Thus, applying this theory, the Hooke's law can be expressed in a new coordinate system as

$$\sigma'_{ij} = C'_{ijkl}\varepsilon'_{kl} \tag{4.21}$$

Now because the stress and strain must meet the transformation laws as they are second-order tensors,

$$\begin{cases} \sigma'_{ij} = Q_{ik}Q_{jl}\sigma_{kl}, & \sigma_{ij} = Q_{ki}Q_{lj}\sigma'_{kl} \\ \varepsilon'_{ij} = Q_{ik}Q_{jl}\varepsilon_{kl}, & \varepsilon_{ij} = Q_{ki}Q_{lj}\varepsilon'_{kl} \end{cases} \tag{4.22}$$

Combining Eq. (4.21) with Eq. (4.22) yields the transformation law for the elasticity tensor

$$C'_{ijkl} = Q_{im}Q_{jn}Q_{kp}Q_{lq}C_{mnpq} \tag{4.23}$$

If the material response is the same under a specific transformation $\boldsymbol{Q}$, Eq. (4.23) reduces to

$$C_{ijkl} = Q_{im}Q_{jn}Q_{kp}Q_{lq}C_{mnpq} \qquad (4.23)$$

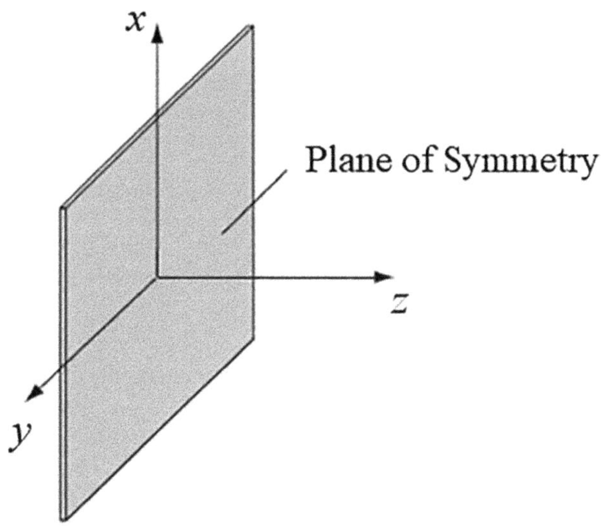

Fig. 4.3 Plane of symmetry for a monoclinic material

This material symmetry relation provides a system of equations that allows reduction in the number of independent elastic constants. We now consider some specific cases of practical interest.

4.4.1.1 Plane of Symmetry (Monoclinic Material)

We first investigate the case of a material with a plane of symmetry. Such a medium is commonly referred to as a monoclinic material. We consider the case of symmetry with respect to the x-y plane as shown in Fig. 4.3.

For this particular symmetry, the required transformation is simply a mirror reflection about the x-y plane and is given by

$$Q_{ij} = \begin{bmatrix} 1 & 0 & 0 \\ 0 & 1 & 0 \\ 0 & 0 & -1 \end{bmatrix} \tag{4.24}$$

Note that this transformation is not a simple rotation that preserves the right-handedness of the coordinate system; that is, it is not a proper orthogonal transformation. Nevertheless, it can be used for our investigations on symmetry. Using this specific transformation in Eq. (4.23) gives $C_{ijkl} = -C_{ijkl}$ if the index 3 appears an odd number of times, and thus these particular elastic constants would have to vanish. In terms of the contracted notation, this gives

$$C_{i4} = C_{i5} = C_{46} = C_{56} = 0, (i = 1, 2, 3) \tag{4.25}$$

Thus, the elastic stiffness matrix takes the form

$$C_{ij} = \begin{bmatrix} C_{11} & C_{12} & C_{13} & 0 & 0 & C_{16} \\ \cdot & C_{22} & C_{23} & 0 & 0 & C_{26} \\ \cdot & \cdot & C_{33} & 0 & 0 & C_{36} \\ \cdot & \cdot & \cdot & C_{44} & C_{45} & 0 \\ \cdot & Sym & \cdot & \cdot & C_{55} & 0 \\ \cdot & \cdot & \cdot & \cdot & \cdot & C_{66} \end{bmatrix} \tag{4.26}$$

It is therefore observed that 13 independent elastic constants are needed to characterize monoclinic materials.

4.4.1.2 Three Perpendicular Planes of Symmetry (Orthotropic Material)

A material with three mutually perpendicular planes of symmetry is called *orthotropic material*. Common examples of such materials include wood and fiber-reinforced composites. To investigate the material symmetries for this case, it is convenient to let the symmetric planes correspond to the coordinate planes shown in Fig. 4.4.

The symmetric relations can be determined by using 180° rotations about each of the coordinate axes. Another convenient scheme is to start with the reduced form from the previous monoclinic case, and re-apply the same transformation with respect to the *y*-*z* plane. This results in the additional elastic constants being reduced to zero:

$$C_{16} = C_{26} = C_{36} = C_{45} = 0 \tag{4.27}$$

Thus, the elastic stiffness matrix for the orthotropic case reduces to having only nine independent elastic constants given by

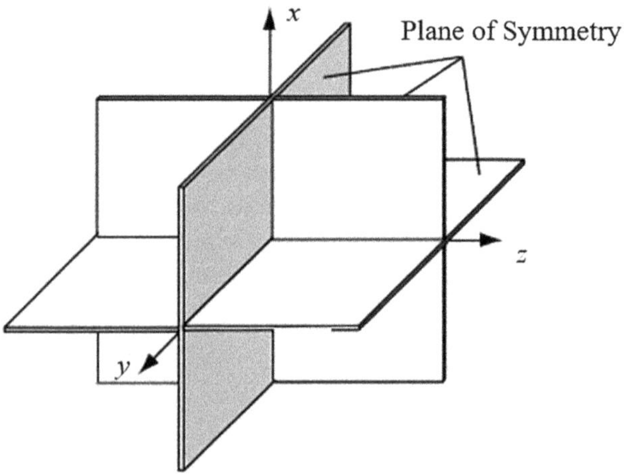

Fig. 4.4 Three symmetric planes for an orthotropic material

$$C_{ij} = \begin{bmatrix} C_{11} & C_{12} & C_{13} & 0 & 0 & 0 \\ \cdot & C_{22} & C_{23} & 0 & 0 & 0 \\ \cdot & \cdot & C_{33} & 0 & 0 & 0 \\ \cdot & \cdot & \cdot & C_{44} & 0 & 0 \\ \cdot & Sym & \cdot & \cdot & C_{55} & 0 \\ \cdot & \cdot & \cdot & \cdot & \cdot & C_{66} \end{bmatrix} \tag{4.28}$$

It should be noted that only two transformations are needed to develop the final reduced constitutive form given in Eq. (4.28). The material also must satisfy a third required transformation that the properties would be the same under a reflection of the x-z plane. However, attempting this transformation would only give relations that are identically satisfied. Thus, for some materials the reduced constitutive form may be developed using only a portion of the total material symmetries. Another issue for orthotropic materials is that the vanish of shear strains implies the vanish of shear stresses, and thus the principal axes of stress coincide with the principal axes of strain. This result is of course not true for general anisotropic materials, for example, the monoclinic constitutive form given in Eq. (4.26).

For orthotropic materials, the elastic compliance matrix has the similar form but is commonly written using notation related to the isotropic theory:

$$S_{ij} = \begin{bmatrix} \frac{1}{E_1} & -\frac{\nu_{21}}{E_2} & -\frac{\nu_{31}}{E_3} & 0 & 0 & 0 \\ -\frac{\nu_{12}}{E_1} & \frac{1}{E_2} & -\frac{\nu_{32}}{E_3} & 0 & 0 & 0 \\ -\frac{\nu_{13}}{E_1} & -\frac{\nu_{23}}{E_2} & \frac{1}{E_3} & 0 & 0 & 0 \\ \cdot & \cdot & \cdot & \frac{1}{\mu_{23}} & 0 & 0 \\ \cdot & Sym & \cdot & \cdot & \frac{1}{\mu_{31}} & 0 \\ \cdot & \cdot & \cdot & \cdot & \cdot & \frac{1}{\mu_{12}} \end{bmatrix} \tag{4.29}$$

where E_i are Young's moduli in the three directions of material symmetry, ν_{ij} are the Poisson's ratios defined by $-e_j/e_i$ for a stress in the i direction, and μ_{ij} are the shear moduli in the $i-j$ planes. Symmetry of this matrix requires that $\nu_{ij}/E_i = \nu_{ji}/E_j$.

4.4.1.3 Axis of Symmetry (Transversely Isotropic Material)

Another common form of material symmetry is with respect to rotations about an axis. This concept can be specified by stating that a material possess an axis of symmetry of order n when the elastic moduli remain unchanged for rotations of $2\pi/n$ radians about the axis. This situation is shown schematically in Fig. 4.5. The hexagonal packing crystalline case shown in Fig. 4.2 has such a symmetry about

Fig. 4.5 Axis of symmetry for a transversely isotropic material

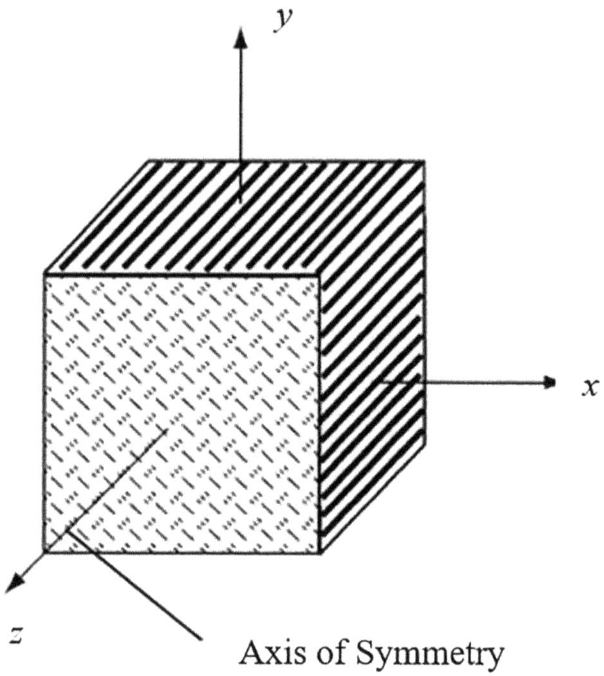

the axis perpendicular to the page for $n = 6$ (60° increments). The only possible symmetries for this case are for orders 2, 3, 4, 6 and infinity. The order 2 case is equivalent to a plane of symmetry previously discussed, and the order 6 is equivalent to the infinite case.

The transformation for arbitrary rotations θ about the z-axis is given by

$$Q_{ij} = \begin{bmatrix} \cos\theta & \sin\theta & 0 \\ -\sin\theta & \cos\theta & 0 \\ 0 & 0 & 1 \end{bmatrix} \tag{4.30}$$

Using this transformation and invoking symmetry for arbitrary rotations corresponds to the case of $n \to \infty$, and such materials are called *transversely isotropic materials*. The elastic stiffness matrix for this case reduces to

$$C_{ij} = \begin{bmatrix} C_{11} & C_{12} & C_{13} & 0 & 0 & 0 \\ \cdot & C_{11} & C_{13} & 0 & 0 & 0 \\ \cdot & \cdot & C_{33} & 0 & 0 & 0 \\ \cdot & \cdot & \cdot & C_{44} & 0 & 0 \\ \cdot & Sym & \cdot & \cdot & C_{44} & 0 \\ \cdot & \cdot & \cdot & \cdot & \cdot & (C_{11} - C_{12})/2 \end{bmatrix} \tag{4.31}$$

Thus, for transversely isotropic materials, only five independent elastic constants exist.

4.4.1.4 Complete Symmetry (Isotropic Material)

For the case of complete symmetry, the material is referred to as an isotropic material, and the fourth-order elasticity tensor has been previously given by

$$C_{ijkl} = \lambda\delta_{ij}\delta_{kl} + \mu\left(\delta_{ik}\delta_{jl} + \delta_{il}\delta_{jk}\right) \tag{4.32}$$

This form can be determined by invoking symmetry with respect to two orthogonal axes, which implies symmetry about the remaining axis. In the contracted matrix form, this result can be expressed as

$$C_{ij} = \begin{bmatrix} \lambda+2\mu & \lambda & \lambda & 0 & 0 & 0 \\ \cdot & \lambda+2\mu & \lambda & 0 & 0 & 0 \\ \cdot & \cdot & \lambda+2\mu & 0 & 0 & 0 \\ \cdot & \cdot & \cdot & \mu & 0 & 0 \\ \cdot & Sym & \cdot & \cdot & \mu & 0 \\ \cdot & \cdot & \cdot & \cdot & \cdot & \mu \end{bmatrix} \tag{4.33}$$

Thus, as shown previously, only two independent elastic constants are present for isotropic materials. For each case presented, a similar elastic compliance matrix can be developed. Our brief presentation does not include all cases of material symmetry. Based on symmetry planes, it has been proven that there are only eight symmetries for linear anisotropic elastic materials for the general case. Of course, a large variety of curvilinear material symmetries also exist in various biological and synthetic materials.

Example 4.1: Hydrostatic Compression of a Monoclinic Cube
Question: Considering a simple example of a cube of monoclinic material under hydrostatic compression, the task is to analyze the differences in mechanical behaviors between isotropic and anisotropic materials.

Solution: In case of hydrostatic compression, the state of stress is given by $\sigma_{ij} = -p\delta_{ij}$, and the monoclinic Hooke's law in compliance form would read as follows:

$$\begin{bmatrix} \varepsilon_x \\ \varepsilon_y \\ \varepsilon_z \\ 2\varepsilon_{yz} \\ 2\varepsilon_{zx} \\ 2\varepsilon_{xy} \end{bmatrix} = \begin{bmatrix} S_{11} & S_{12} & S_{13} & 0 & 0 & S_{16} \\ \cdot & S_{22} & S_{23} & 0 & 0 & S_{26} \\ \cdot & \cdot & S_{33} & 0 & 0 & S_{36} \\ \cdot & \cdot & \cdot & S_{44} & S_{45} & 0 \\ \cdot & sym & \cdot & \cdot & S_{55} & 0 \\ \cdot & \cdot & \cdot & \cdot & \cdot & S_{66} \end{bmatrix} \begin{bmatrix} -p \\ -p \\ -p \\ 0 \\ 0 \\ 0 \end{bmatrix} \tag{4.34}$$

Expanding the matrix relation (Eq. (4.34)) gives the following deformation field components:

$$\begin{cases} \varepsilon_x = -(S_{11} + S_{12} + S_{13})p \\ \varepsilon_y = -(S_{12} + S_{22} + S_{23})p \\ \varepsilon_z = -(S_{13} + S_{23} + S_{33})p \\ \varepsilon_{yz} = 0 \\ \varepsilon_{zx} = 0 \\ \varepsilon_{xy} = -\frac{1}{2}(S_{16} + S_{26} + S_{36})p \end{cases} \tag{4.35}$$

The corresponding strains for the isotropic case would be given by

$$\begin{cases} \varepsilon_x = \varepsilon_y = \varepsilon_z = -[(1-2\nu)/E]p \\ \varepsilon_{yz} = \varepsilon_{zx} = \varepsilon_{xy} = 0 \end{cases} \tag{4.36}$$

Thus, the response of the monoclinic material is considerably different from isotropic behavior and yields a nonzero shear strain even under uniform hydrostatic stress. Additional examples using simple shear and/or bending deformations can also be used to demonstrate the complexity of anisotropic stress-strain behavior.

4.4.2 Independent Elastic Constants for Elastic Solids

In this part, the number of independent elastic constants for different elastic solids, starting from the most complex elastic solids to the simplest elastic solids, and the process to obtain the number of independent elastic constants are discussed.

4.4.2.1 Fully Anisotropic Elastic Solids

The fully anisotropic elastic solid is the most complex one. In this solid, no symmetry is present. However, it is known from the previous part that the elastic stiffness matrix is always symmetric. Therefore, for the most complex elastic solid, i.e., the fully anisotropic elastic solid, there are 21 independent elastic constants in the matrix. The elastic stiffness matrix can be written as:

$$C_{mn} = C_{nm} = \begin{bmatrix} C_{11} & C_{12} & C_{13} & C_{14} & C_{15} & C_{16} \\ & C_{22} & C_{23} & C_{24} & C_{25} & C_{26} \\ \cdot & \cdot & C_{33} & C_{34} & C_{35} & C_{36} \\ \cdot & \cdot & \cdot & C_{44} & C_{45} & C_{46} \\ \cdot & Sym & \cdot & \cdot & C_{55} & C_{56} \\ & \cdot & \cdot & \cdot & \cdot & C_{66} \end{bmatrix} \tag{4.37}$$

To determine these material constants, the mechanical testing needs to be performed. 21 independent constants in the elasticity matrix implies that 21 independent mechanical testing might be needed, which is not feasible in the design of

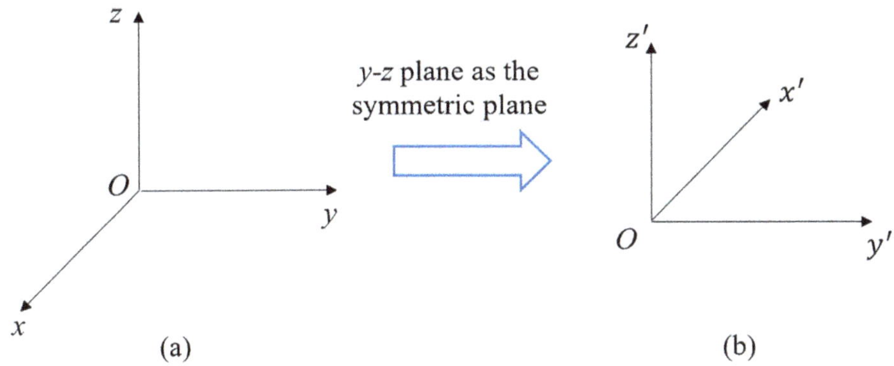

Fig. 4.6 Illustration of the symmetric plane (*y*-*z* plane) for anisotropic elastic solids: (**a**) Old coordinate system; (**b**) New coordinate system

Table 4.1 Relation between the old and new coordinate systems

	x	y	z
x'	$l_1 = -1$	$m_1 = 0$	$n_1 = 0$
y'	$l_2 = 0$	$m_2 = 1$	$n_2 = 0$
z'	$l_3 = 0$	$m_3 = 0$	$n_3 = 1$

mechanical experiments. Therefore, the fully anisotropic elastic solid is seldom used to describe the mechanical behavior of the engineering materials. Simplifications and assumptions need to be made to reduce the number of independent elastic constants in the elasticity matrix.

4.4.2.2 Anisotropic Elastic Solids with One Symmetric Plane

This kind of solids has one symmetric plane. The *y*-*z* plane is supposed to be the symmetric plane (Fig. 4.6). The relation between the old ($Oxyz$) and new ($Ox'y'z'$) systems can be written in Table 4.1:

In the table, l_i, m_i, n_i are the cosines from $Ox'y'z'$ to $Oxyz$ system. Because the *y*-*z* plane is the symmetric plane, the stress calculated at the same point in the two systems should be equal to each other and based on the stress transformation relation given in Eq. (3.18), we have:

$$\begin{cases} \sigma_{x'} = \sigma_x, \sigma_{y'} = \sigma_y, \sigma_{z'} = \sigma_z \\ \tau_{y'z'} = \tau_{yz}, \tau_{x'z'} = -\tau_{xz}, \tau_{x'y'} = -\tau_{xy} \end{cases} \tag{4.38}$$

Similarly, the strain calculated at the same point in the two systems should also be equal to each other and based on the strain transformation relation given in Eq. (2.32), we have:

$$\begin{cases} \varepsilon_{x'} = \varepsilon_x, \varepsilon_{y'} = \varepsilon_y, \varepsilon_{z'} = \varepsilon_z \\ \gamma_{y'z'} = \gamma_{yz}, \gamma_{x'z'} = -\gamma_{xz}, \gamma_{x'y'} = -\gamma_{xy} \end{cases} \quad (4.39)$$

Putting Eqs. (4.38) and (4.39) into the constitutive relation given in Eq. (4.1), the stress-strain relation in the $Ox'y'z'$ system can be written as:

$$\begin{cases} \sigma_{x'} = C_{11}\varepsilon_{x'} + C_{12}\varepsilon_{y'} + C_{13}\varepsilon_{z'} + C_{14}\gamma_{y'z'} - C_{15}\gamma_{x'z'} - C_{16}\gamma_{x'y'} \\ \sigma_{y'} = C_{21}\varepsilon_{x'} + C_{22}\varepsilon_{y'} + C_{23}\varepsilon_{z'} + C_{24}\gamma_{y'z'} - C_{25}\gamma_{x'z'} - C_{26}\gamma_{x'y'} \\ \sigma_{z'} = C_{31}\varepsilon_{x'} + C_{32}\varepsilon_{y'} + C_{33}\varepsilon_{z'} + C_{34}\gamma_{y'z'} - C_{35}\gamma_{x'z'} - C_{36}\gamma_{x'y'} \\ \tau_{y'z'} = C_{41}\varepsilon_{x'} + C_{42}\varepsilon_{y'} + C_{43}\varepsilon_{z'} + C_{44}\gamma_{y'z'} - C_{45}\gamma_{x'z'} - C_{46}\gamma_{x'y'} \\ -\tau_{x'z'} = C_{51}\varepsilon_{x'} + C_{52}\varepsilon_{y'} + C_{43}\varepsilon_{z'} + C_{54}\gamma_{y'z'} - C_{55}\gamma_{x'z'} - C_{56}\gamma_{x'y'} \\ -\tau_{x'y} = C_{61}\varepsilon_{x'} + C_{62}\varepsilon_{y'} + C_{63}\varepsilon_{z'} + C_{64}\gamma_{y'z'} - C_{65}\gamma_{x'z'} - C_{66}\gamma_{x'y'} \end{cases} \quad (4.40)$$

The constitutive relations, setting up in the two systems, i.e., Eqs. (4.1) and (4.40), are to describe the same solid, and thus should be equal to each other. Therefore, we have:

$$C_{15} = C_{16} = C_{25} = C_{26} = C_{35} = C_{36} = C_{45} = C_{46} = 0 \quad (4.41)$$

Therefore, the elastic stiffness matrix can be written in the reduced form as below:

$$C_{ij} = \begin{bmatrix} C_{11} & C_{12} & C_{13} & C_{14} & 0 & 0 \\ & C_{22} & C_{23} & C_{24} & 0 & 0 \\ \cdot & \cdot & C_{33} & C_{34} & 0 & 0 \\ \cdot & \cdot & \cdot & C_{44} & 0 & 0 \\ \cdot & Sym & \cdot & \cdot & C_{55} & C_{56} \\ & \cdot & \cdot & \cdot & \cdot & C_{66} \end{bmatrix} \quad (4.42)$$

It can be seen that there are 13 independent elastic constants in the stiffness matrix for the anisotropic elastic solids with one symmetric plane.

4.4.2.3 Orthotropic Elastic Solids

The orthotropic elastic solid has two symmetric planes. The *y-z* and *x-z* planes are assumed to be the symmetric planes (Fig. 4.7). From the previous part, it is known that when the *y-z* plane is the symmetric plane, the 21 independent elastic constants can be reduced to 13. Now let's use the condition that the *x-z* plane is the symmetric plane to further simplify the elastic stiffness matrix (Fig. 4.7b).

When *x-z* plane is set as the symmetric plane, the relation between the old ($Oxyz$) and new ($Ox'y'z'$) systems can be written in Table 4.2:

In the table, l_i, m_i, n_i are the cosines from $Ox'y'z'$ to $Oxyz$ system (Fig. 4.7b). Because *x-z* plane is the symmetric plane, the stress calculated at the same point in

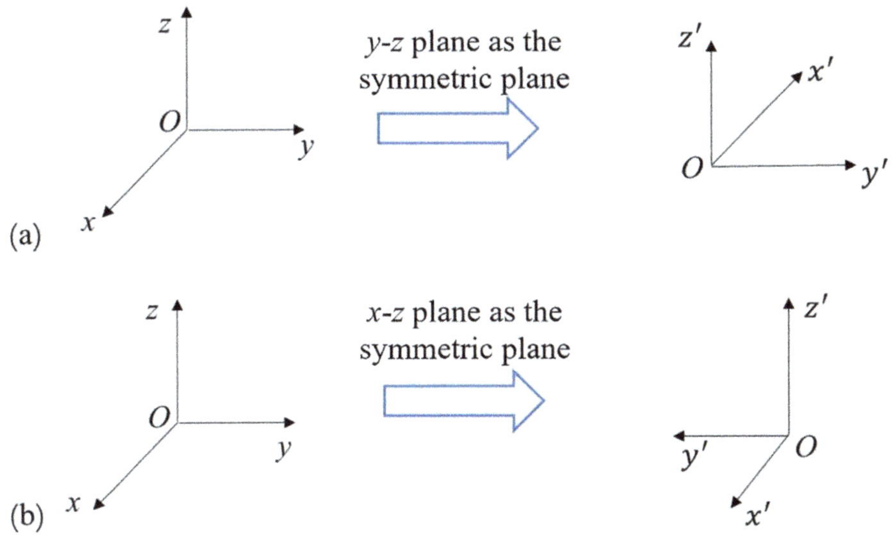

Fig. 4.7 Illustration of the two symmetric planes for orthotropic elastic solids: (**a**) *y*-*z* plane as the symmetric plane; (**b**) *x*-*z* plane as the symmetric plane

Table 4.2 Relation between the old and new coordinate systems in Fig. 4.7b

	x	y	z
x'	$l_1 = 1$	$m_1 = 0$	$n_1 = 0$
y'	$l_2 = 0$	$m_2 = -1$	$n_2 = 0$
z'	$l_3 = 0$	$m_3 = 0$	$n_3 = 1$

the two systems should be equal to each other and based on the stress transformation relation given in Eq. (3.18), we have:

$$\begin{cases} \sigma_{x'} = \sigma_x, \sigma_{y'} = \sigma_y, \sigma_{z'} = \sigma_z \\ \tau_{y'z'} = -\tau_{yz}, \tau_{x'z'} = \tau_{xz}, \tau_{x'y'} = -\tau_{xy} \end{cases} \tag{4.43}$$

Similarly, the strain calculated at the same point in the two systems should also be equal to each other and based on the strain transformation relation given in Eq. (2.32), we have:

$$\begin{cases} \varepsilon_{x'} = \varepsilon_x, \varepsilon_{y'} = \varepsilon_y, \varepsilon_{z'} = \varepsilon_z \\ \gamma_{y'z'} = -\gamma_{yz}, \gamma_{x'z'} = \gamma_{xz}, \gamma_{x'y'} = -\gamma_{xy} \end{cases} \tag{4.44}$$

Putting Eqs. (4.43) and (4.44) into the constitutive relation given in Eq. (4.1), the stress-strain relation in the $Ox'y'z'$ system can be written as:

$$\begin{cases} \sigma_{x'} = C_{11}\varepsilon_{x'} + C_{12}\varepsilon_{y'} + C_{13}\varepsilon_{z'} - C_{14}\gamma_{y'z'} + C_{15}\gamma_{x'z'} - C_{16}\gamma_{x'y'} \\ \sigma_{y'} = C_{21}\varepsilon_{x'} + C_{22}\varepsilon_{y'} + C_{23}\varepsilon_{z'} - C_{24}\gamma_{y'z'} + C_{25}\gamma_{x'z'} - C_{26}\gamma_{x'y'} \\ \sigma_{z'} = C_{31}\varepsilon_{x'} + C_{32}\varepsilon_{y'} + C_{33}\varepsilon_{z'} - C_{34}\gamma_{y'z'} + C_{35}\gamma_{x'z'} - C_{36}\gamma_{x'y'} \\ -\tau_{y'z'} = C_{41}\varepsilon_{x'} + C_{42}\varepsilon_{y'} + C_{43}\varepsilon_{z'} - C_{44}\gamma_{y'z'} + C_{45}\gamma_{x'z'} - C_{46}\gamma_{x'y'} \\ \tau_{x'z'} = C_{51}\varepsilon_{x'} + C_{52}\varepsilon_{y'} + C_{43}\varepsilon_{z'} - C_{54}\gamma_{y'z'} + C_{55}\gamma_{x'z'} - C_{56}\gamma_{x'y'} \\ -\tau_{x'y'} = C_{61}\varepsilon_{x'} + C_{62}\varepsilon_{y'} + C_{63}\varepsilon_{z'} - C_{64}\gamma_{y'z'} + C_{65}\gamma_{x'z'} - C_{66}\gamma_{x'y'} \end{cases} \tag{4.45}$$

The constitutive relations, setting up in the two systems, i.e., Eqs. (4.1) and (4.45), are to describe the same solid, and thus should be equal to each other, and thus we further have:

$$C_{14} = C_{24} = C_{34} = C_{56} = 0 \tag{4.46}$$

Therefore, the elastic stiffness matrix given in Eq. (4.42) can be further reduced to the following form:

$$C_{ij} = \begin{bmatrix} C_{11} & C_{12} & C_{13} & 0 & 0 & 0 \\ & C_{22} & C_{23} & 0 & 0 & 0 \\ \cdot & \cdot & C_{33} & 0 & 0 & 0 \\ \cdot & \cdot & \cdot & C_{44} & 0 & 0 \\ \cdot & Sym & \cdot & \cdot & C_{55} & 0 \\ & \cdot & \cdot & \cdot & \cdot & C_{66} \end{bmatrix} \tag{4.47}$$

It can be seen four more constants can be eliminated from the stiffness matrix. Therefore, there are nine independent elastic constants in the elastic stiffness matrix for the orthotropic elastic solids. It should be noted that if *x*-*y* plane is set as another symmetric plane, no new reduction can be obtained implying that if two orthotropic planes are the symmetric planes, the third one must be the symmetric one as well. The reduced form stiffness matrix implies that the normal and shear stresses (strains) are decoupled. For example, the normal stresses only induce the normal strains and the shear stresses only induce the shear strains.

4.4.2.4 Transversely Isotropic Elastic Solids

Two symmetric planes and one isotropic plane are present in the transversely isotropic elastic solid, supposing the *y*-*z* and *x*-*z* planes are the symmetric planes and the *x*-*y* plane is the isotropic plane (Fig. 4.8). The process to simplify the stiffness matrix using the condition of two symmetric planes is the same as that for the orthotropic elastic solid. Now we only need to use the condition of the isotropic plane to further simplify the stiffness matrix. Because *x*-*y* plane is isotropic, we rotate the plane along *z* axis by any degree and the same constitutive relation should be obtained. Below we simplify the stiffness matrix in two steps.

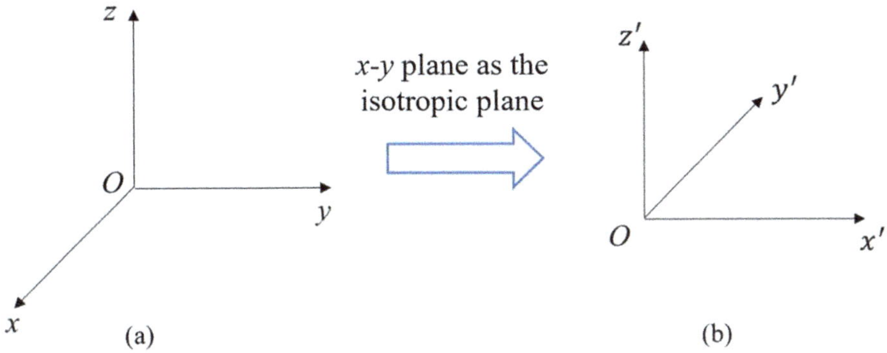

Fig. 4.8 Illustration of the isotropic planes for transversely isotropic elastic solids: (**a**) old coordinate system; (**b**) new coordinate system

Table 4.3 Relation between the old and new coordinate systems

	x	y	z
x'	$l_1 = 0$	$m_1 = 1$	$n_1 = 0$
y'	$l_2 = -1$	$m_2 = 0$	$n_2 = 0$
z'	$l_3 = 0$	$m_3 = 0$	$n_3 = 1$

First, we rotate the x-y plane along z axis by 90°. The relation between the old and new coordinate systems can be written in Table 4.3.

In the table, l_i, m_i, n_i are the cosines from $Ox'y'z'$ to $Oxyz$system (Fig. 4.8). The stress calculated at the same point in the two coordinate systems should be equal to each other and based on the stress transformation relation given in Eq. (3.18), we have:

$$\begin{cases} \sigma_{x'} = \sigma_y, \sigma_{y'} = \sigma_x, \sigma_{z'} = \sigma_z \\ \tau_{x'z'} = \tau_{yz}, \tau_{y'z'} = -\tau_{xz}, \tau_{y'x'} = -\tau_{xy} \end{cases} \tag{4.48}$$

Similarly, the strain calculated at the same point in the two coordinate systems should also be equal to each other and based on the strain transformation relation given in Eq. (2.32), we have:

$$\begin{cases} \varepsilon_{x'} = \varepsilon_y, \varepsilon_{y'} = \varepsilon_x, \varepsilon_{z'} = \varepsilon_z \\ \gamma_{x'z'} = \gamma_{yz}, \gamma_{y'z'} = -\gamma_{xz}, \gamma_{y'x'} = -\gamma_{xy} \end{cases} \tag{4.49}$$

Putting Equations. (4.48) and (4.49) into the constitutive relation given in Eq. (4.1), the stress-strain relation in the $Ox'y'z'$ system can be written as:

$$\begin{cases} \sigma_{y'} = C_{11}\varepsilon_{y'} + C_{12}\varepsilon_{x'} + C_{13}\varepsilon_{z'} \\ \sigma_{x'} = C_{12}\varepsilon_{y'} + C_{22}\varepsilon_{x'} + C_{23}\varepsilon_{z'} \\ \sigma_{z'} = C_{13}\varepsilon_{y'} + C_{23}\varepsilon_{x'} + C_{33}\varepsilon_{z'} \\ \tau_{x'z'} = C_{44}\gamma_{x'z'} \\ -\tau_{y'z'} = -C_{55}\gamma_{y'z'} \\ -\tau_{x'y'} = -C_{66}\gamma_{x'y'} \end{cases} \tag{4.50}$$

The constitutive relations, setting up in the two coordinate systems, i.e., Eqs. (4.1) and (4.50), are to describe the same solid, and thus should equal, thus we have:

$$C_{11} = C_{22}, C_{13} = C_{23}, C_{44} = C_{55} \tag{4.51}$$

Therefore, the elastic stiffness matrix can be written in the reduced form below:

$$C_{ij} = \begin{bmatrix} C_{11} & C_{12} & C_{13} & 0 & 0 & 0 \\ & C_{11} & C_{13} & 0 & 0 & 0 \\ \cdot & \cdot & C_{33} & 0 & 0 & 0 \\ \cdot & \cdot & \cdot & C_{44} & 0 & 0 \\ \cdot & Sym & \cdot & \cdot & C_{44} & 0 \\ & \cdot & \cdot & \cdot & \cdot & C_{66} \end{bmatrix} \tag{4.52}$$

It can be seen the number of independent elastic constant has been reduced to 6. We can further reduce it using the second condition as below:

Second, we rotate the *x-y* plane along *z* axis by alpha degrees (Fig. 4.9). The relation between the old and new coordinate systems can be written in Table 4.4.

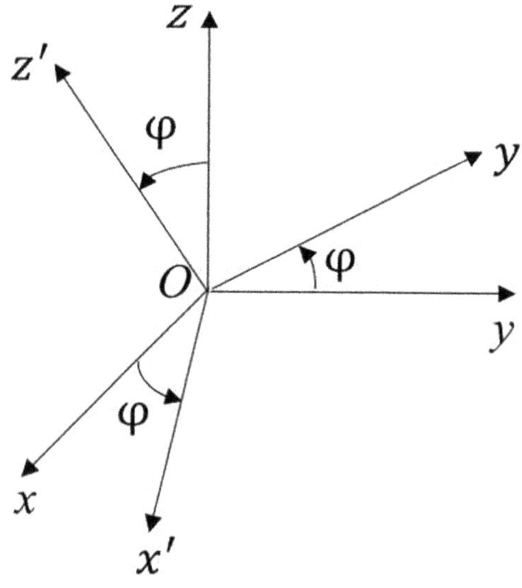

Fig. 4.9 Rotation of the *x-y* plane by alpha degrees

Table 4.4 Relation between the old and new coordinate systems

	x	y	z
x'	$l_1 = \cos \varphi$	$m_1 = \sin \varphi$	$n_1 = 0$
y'	$l_2 = -\sin \varphi$	$m_2 = \cos \varphi$	$n_2 = 0$
z'	$l_3 = 0$	$m_3 = 0$	$n_3 = 1$

In the table, l_i, m_i, n_i are the cosines from $Ox'y'z'$ to $Oxyz$system (Fig. 4.9). Then based on the stress transformation relation given in Eq. (3.18) and the strain transformation relation given in Eq. (2.32), we have:

$$\begin{cases} \tau_{x'y'} = \frac{1}{2}(\sigma_y - \sigma_x)\sin 2\varphi + \tau_{xy}\cos 2\varphi \\ \gamma_{x'y'} = (\varepsilon_y - \varepsilon_x)\sin 2\varphi + \gamma_{xy}\cos 2\varphi \end{cases} \tag{4.53}$$

On the other hand, because $\tau_{y'x'} = -\tau_{xy}$ and $\gamma_{y'x'} = -\gamma_{xy}$, from relation (4.1), we have:

$$\tau_{x'y'} = C_{66}\gamma_{x'y'} \tag{4.54}$$

Putting Eq. (4.53) into Eq. (4.54), we have

$$\frac{1}{2}(\sigma_y - \sigma_x)\sin 2\varphi + \tau_{xy}\cos 2\varphi = C_{66}\left[(\varepsilon_y - \varepsilon_x)\sin 2\varphi + \gamma_{xy}\cos 2\varphi\right] \tag{4.55}$$

Putting $\tau_{xy} = C_{66}\gamma_{xy}$ into Eq. (4.55), we can obtain:

$$\sigma_y - \sigma_x = 2C_{66}(\varepsilon_y - \varepsilon_x) \tag{4.56a}$$

On the other hand, from Eq. (4.1), we have:

$$\sigma_y - \sigma_x = (C_{11} - C_{12})(\varepsilon_y - \varepsilon_x) \tag{4.56b}$$

Comparing Eq. (4.56a) with Eq. (4.56b), we have:

$$2C_{66} = (C_{11} - C_{12})$$

Therefore, one more independent elastic constant can be eliminated and there are only 5 independent elastic constants in the stiffness matrix for the transversely isotropic elastic solid, which can be written as below:

$$C_{ij} = \begin{bmatrix} C_{11} & C_{12} & C_{13} & 0 & 0 & 0 \\ & C_{11} & C_{13} & 0 & 0 & 0 \\ \cdot & \cdot & C_{33} & 0 & 0 & 0 \\ \cdot & \cdot & \cdot & C_{44} & 0 & 0 \\ \cdot & Sym & \cdot & \cdot & C_{44} & 0 \\ & \cdot & \cdot & \cdot & \cdot & (C_{11} - C_{12})/2 \end{bmatrix} \tag{4.57}$$

4.4.2.5 Linear Isotropic Elastic Solids

In the linear isotropic elastic solid, the mechanical properties in any direction of the solid are the same. The stiffness matrix of the linear isotropic elastic solid can be obtained by adding one more transformation condition onto those of the transversely isotropic elastic solid, i.e., in addition to the conditions in the previous part, the following condition should be used: the constitutive relation stays the same after rotating the *y*-*z* plane along *x* axis by 90 degrees (Fig. 4.10).

The stress tensors calculated from the two coordinate systems equal and based on the stress transformation relation given in Eq. (3.18), we have:

$$\begin{cases} \sigma_{x'} = \sigma_x, \sigma_{y'} = \sigma_z, \sigma_{z'} = \sigma_y \\ \tau_{y'z'} = -\tau_{yz}, \tau_{x'z'} = -\tau_{xy}, \tau_{x'y'} = \tau_{xz} \end{cases} \tag{4.58}$$

Furthermore, the strain tensors calculated from the two systems equal and based on the strain transformation relation given in Eq. (2.32), we have:

$$\begin{cases} \varepsilon_{x'} = \varepsilon_x, \varepsilon_{y'} = \varepsilon_z, \varepsilon_{z'} = \varepsilon_y \\ \gamma_{y'z'} = -\gamma_{yz}, \gamma_{x'z'} = -\gamma_{xy}, \gamma_{x'y'} = \gamma_{xz} \end{cases} \tag{4.59}$$

Putting the relations given in Eqs. (4.58) and (4.59) into constitutive relation (Eq. (4.1)), the stress-strain relation in the $Ox'y'z'$ system can be written as:

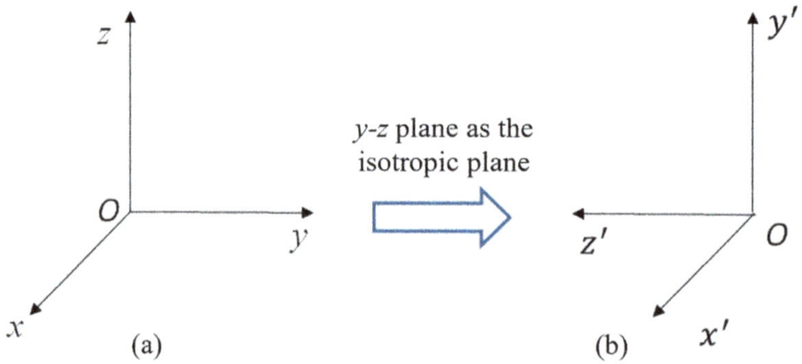

Fig. 4.10 Rotation of the *y*-*z* plane by 90°: (**a**) old coordinate system; (**b**) new coordinate system

$$\begin{cases} \sigma_{x'} = C_{11}\varepsilon_{x'} + C_{12}\varepsilon_{z'} + C_{13}\varepsilon_{y'} \\ \sigma_{y'} = C_{12}\varepsilon_{x'} + C_{11}\varepsilon_{z'} + C_{13}\varepsilon_{y'} \\ \sigma_{z'} = C_{13}\varepsilon_{x'} + C_{13}\varepsilon_{z'} + C_{33}\varepsilon_{y'} \\ -\tau_{y'z'} = -C_{44}\gamma_{y'z'} \\ \tau_{x'y'} = C_{44}\gamma_{x'y'} \\ -\tau_{x'z'} = -\frac{1}{2}(C_{11} - C_{12})\gamma_{x'y'} \end{cases} \tag{4.60}$$

The constitutive relations, setting up in the two systems, i.e., Eq. (4.1) and (4.60), are to describe the same solid, and thus should equal, thus we have:

$$C_{12} = C_{13}, C_{11} = C_{33}, C_{44} = (C_{11} - C_{12})/2 \tag{4.61}$$

Therefore, the stiffness matrix can be written in the reduced form as below:

$$C_{ij} = \begin{bmatrix} C_{11} & C_{12} & C_{12} & 0 & 0 & 0 \\ & C_{11} & C_{12} & 0 & 0 & 0 \\ \cdot & \cdot & C_{11} & 0 & 0 & 0 \\ \cdot & \cdot & \cdot & (C_{11} - C_{12})/2 & 0 & 0 \\ \cdot & Sym & \cdot & \cdot & (C_{11} - C_{12})/2 & 0 \\ & \cdot & \cdot & \cdot & \cdot & (C_{11} - C_{12})/2 \end{bmatrix} \tag{4.62}$$

It can be seen there are only 2 independent elastic constants in the stiffness matrix for the linear isotropic elastic solid.

4.5 Constitutive Relation for Isotropic Elastic Solids

From the previous part, only two independent elastic constants are required to describe the mechanical behavior of the isotropic elastic solid. Its constitutive relation can be written in the index format as below:

$$\sigma_{ij} = C_{ijkl}\varepsilon_{kl} \tag{4.63}$$

Using the kinematic rule, the second-order stress and strain tensors can be reduced to the first-order tensors and their relationship can be written as below:

$$\begin{cases} \sigma_x = \lambda(\varepsilon_x + \varepsilon_y + \varepsilon_z) + 2\mu\varepsilon_x \\ \sigma_y = \lambda(\varepsilon_x + \varepsilon_y + \varepsilon_z) + 2\mu\varepsilon_y \\ \sigma_z = \lambda(\varepsilon_x + \varepsilon_y + \varepsilon_z) + 2\mu\varepsilon_z \\ \tau_{xy} = \mu\gamma_{xy} \\ \tau_{yz} = \mu\gamma_{yz} \\ \tau_{zx} = \mu\gamma_{zx} \end{cases} \tag{4.64}$$

The equations can be written in terms of the index format as:

$$\sigma_{ij} = \lambda\varepsilon_{kk}\delta_{ij} + 2\mu\varepsilon_{ij} \tag{4.65}$$

where, $\lambda = C_{12}$ is the Lame's constant, $\mu = (C_{11} - C_{12})/2$ is the shear modulus or the modulus of rigidity. The constitutive relation, i.e., Eq. (4.64), is to use the strain components to express the stress components. The inverted form, i.e., use the stress components to express the strain components, is also called the *constitutive relation*. The inverted form of constitutive relation takes the following form:

$$\begin{cases} \varepsilon_x = \frac{1}{E}\left[\sigma_x - \nu(\sigma_y + \sigma_z)\right] \\ \varepsilon_y = \frac{1}{E}\left[\sigma_y - \nu(\sigma_z + \sigma_x)\right] \\ \varepsilon_z = \frac{1}{E}\left[\sigma_z - \nu(\sigma_x + \sigma_y)\right] \\ \frac{1}{2}\gamma_{xy} = \frac{(1+\nu)}{E}\tau_{xy} = \frac{1}{2\mu}\tau_{xy} \\ \frac{1}{2}\gamma_{yz} = \frac{(1+\nu)}{E}\tau_{yz} = \frac{1}{2\mu}\tau_{yz} \\ \frac{1}{2}\gamma_{zx} = \frac{(1+\nu)}{E}\tau_{zx} = \frac{1}{2\mu}\tau_{zx} \end{cases} \tag{4.66}$$

i.e.,

$$\varepsilon_{ij} = \frac{1+\nu}{E}\sigma_{ij} - \frac{\nu}{E}\sigma_{kk}\delta_{ij} \tag{4.67}$$

where,

$$E = \frac{\mu(3\lambda + 2\mu)}{\lambda + \mu} \tag{4.68}$$

is the Young's modulus or the modulus of elasticity,

$$\nu = \frac{\lambda}{2(\lambda + \mu)} \tag{4.69}$$

is the Poisson's ratio.

Equation (4.64) can also be written in the index format as below:

$$\sigma_i = C_{ij}\varepsilon_j, (i,j = 1, 2, \ldots, 6) \tag{4.70}$$

where C_{ij} is called the *elastic stiffness matrix*, taking the following form:

$$C_{ij} = \frac{E}{(1-2\nu)(1+\nu)} \begin{bmatrix} 1-\nu & \nu & \nu & 0 & 0 & 0 \\ \nu & 1-\nu & \nu & 0 & 0 & 0 \\ \nu & \nu & 1-\nu & 0 & 0 & 0 \\ 0 & 0 & 0 & (1-2\nu) & 0 & 0 \\ 0 & 0 & 0 & 0 & (1-2\nu) & 0 \\ 0 & 0 & 0 & 0 & 0 & (1-2\nu) \end{bmatrix} \tag{4.71}$$

Equation (4.66) can also be written as:

$$\varepsilon_i = S_{ij}\sigma_j, (i,j = 1, 2, \ldots, 6) \tag{4.72}$$

where S_{ij} is called the *elastic compliance matrix*, taking the following form:

$$S_{ij} = \begin{bmatrix} 1/E & -\nu/E & -\nu/E & 0 & 0 & 0 \\ -\nu/E & 1/E & -\nu/E & 0 & 0 & 0 \\ -\nu/E & -\nu/E & 1/E & 0 & 0 & 0 \\ 0 & 0 & 0 & \frac{(1+\nu)}{E} & 0 & 0 \\ 0 & 0 & 0 & 0 & \frac{(1+\nu)}{E} & 0 \\ 0 & 0 & 0 & 0 & 0 & \frac{(1+\nu)}{E} \end{bmatrix} \tag{4.73}$$

where the stress and strain take the following forms:

$$\begin{cases} \sigma_i = \left[\sigma_x, \sigma_y, \sigma_z, \sigma_{yz}, \sigma_{xz}, \sigma_{xy}\right]^{\mathrm{T}} \\ \varepsilon_j = \left[\varepsilon_x, \varepsilon_y, \varepsilon_z, \varepsilon_{yz}, \varepsilon_{xz}, \varepsilon_{xy}\right]^{\mathrm{T}} \end{cases} \tag{4.74}$$

4.6 Physical Meaning of Elastic Constants

To better understand the elastic constants of the linear elastic isotropic solid, the physical meanings of these elastic constants are introduced in this part, which are derived using three independent loading scenarios, i.e., simple tension, pure shear and hydrostatic compression.

4.6.1 Simple Tension

From the loading scenario of simple tension, the physical meaning of Young's modulus can be obtained. A sample subjected to a uniaxial tension in the x direction, as shown in Fig. 4.11, is considered. The state of stress is closely represented by the 1D field

$$\sigma_{ij} = \begin{bmatrix} \sigma & 0 & 0 \\ 0 & 0 & 0 \\ 0 & 0 & 0 \end{bmatrix} \tag{4.75}$$

Using this in the strain-stress relation gives the corresponding strain field.

$$\varepsilon_{ij} = \begin{bmatrix} \frac{\sigma}{E} & 0 & 0 \\ 0 & -\frac{\nu}{E}\sigma & 0 \\ 0 & 0 & -\frac{\nu}{E}\sigma \end{bmatrix} \tag{4.76}$$

Therefore, $E = \sigma/\varepsilon_x$ and is simply the slope of the stress-strain curve, while $\nu = -\varepsilon_y/\varepsilon_x = -\varepsilon_z/\varepsilon_x$ is the ratio of the transverse strain to the axial strain. Standard measurement systems can easily collect axial stress, transverse and axial strain data, and thus through this one type of test both elastic constants can be determined for materials of interest.

4.6.2 Pure Shear

From the loading scenario of pure shear, the physical meaning of shear modulus can be obtained. A thin-walled cylinder subjected to torsional loading, as shown in Fig. 4.12, is considered. The state of stress on the surface of the cylindrical sample is given by

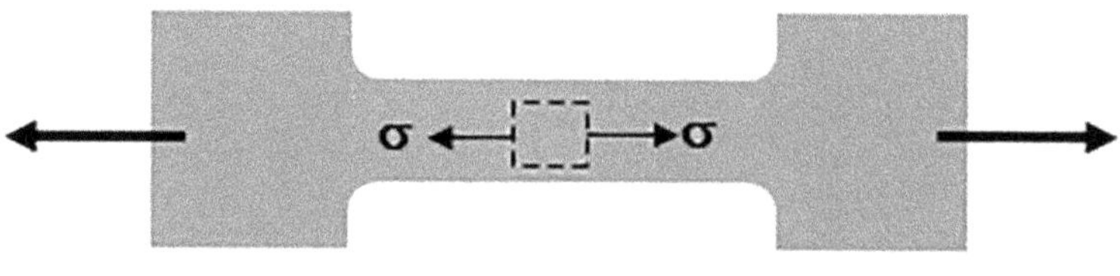

Fig. 4.11 Loading scenario of simple tension

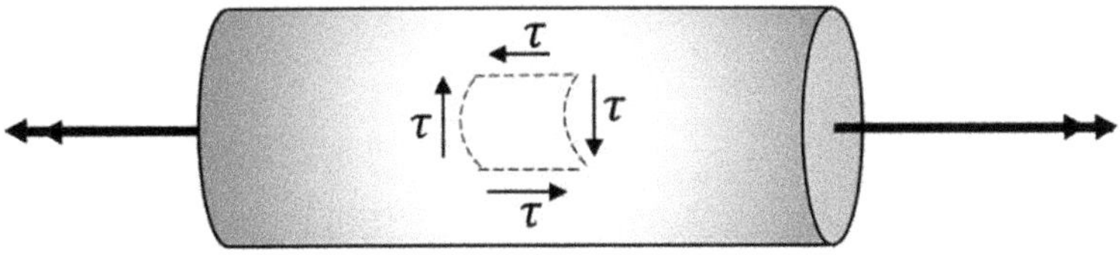

Fig. 4.12 Loading scenario of pure shear

$$\sigma_{ij} = \begin{bmatrix} 0 & \tau & 0 \\ \tau & 0 & 0 \\ 0 & 0 & 0 \end{bmatrix} \tag{4.77}$$

Again, using Hooke's law, the corresponding strain field becomes

$$\varepsilon_{ij} = \begin{bmatrix} 0 & \frac{\tau}{2\mu} & 0 \\ \frac{\tau}{2\mu} & 0 & 0 \\ 0 & 0 & 0 \end{bmatrix} \tag{4.78}$$

and thus, the shear modulus is given by $\mu = \tau/2\varepsilon_{xy} = \tau/\gamma_{xy}$, and this modulus is simply the slope of the shear stress-shear strain curve.

4.6.3 *Hydrostatic compression*

From the loading scenario of pure shear, the physical meaning of the bulk modulus can be obtained. A cube subjected to the uniform compression, as shown in Fig. 4.13, is considered. This type of loading would be realizable if the sample was placed in a high-pressure compression chamber. The state of stress for this case is given by:

$$\sigma_{ij} = \begin{bmatrix} -p & 0 & 0 \\ 0 & -p & 0 \\ 0 & 0 & -p \end{bmatrix} = -p\delta_{ij} \tag{4.79}$$

This is an isotropic state of stress and the strains follow the Hooke's law

$$\varepsilon_{ij} = \begin{bmatrix} -\frac{1-2\nu}{E}p & 0 & 0 \\ 0 & -\frac{1-2\nu}{E}p & 0 \\ 0 & 0 & -\frac{1-2\nu}{E}p \end{bmatrix} \tag{4.80}$$

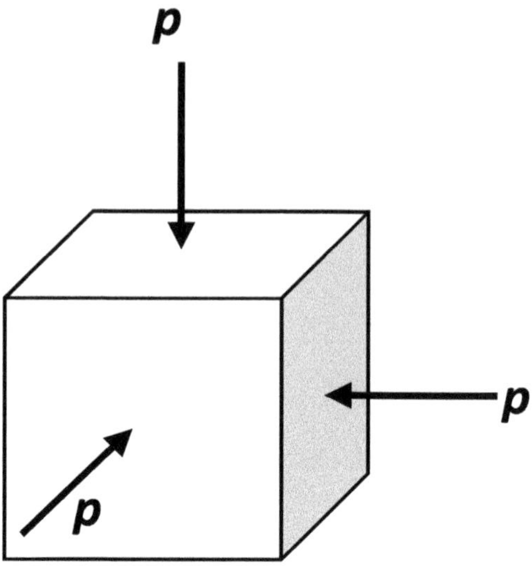

Fig. 4.13 Loading scenario of hydrostatic compression

The dilatation that represents the change in material volume is thus given by $\varepsilon_{kk} = -3(1 - 2\nu)p/E$, which can be written as

$$p = -k \cdot \vartheta \tag{4.81}$$

where $k = E/[3(1 - 2\nu)]$ is called the *bulk modulus of elasticity*. This additional elastic constant represents the ratio of pressure to the dilatation, which could be referred to as the volumetric stiffness of the material. Notice that as Poisson's ratio approaches to 0.5, the bulk modulus becomes unbounded and the material does not undergo any volumetric deformation and hence is incompressible.

4.7 Relationships Between Elastic Constants

Our discussion of elastic constants for isotropic materials has led to the definition of five elastic constants λ, μ, E, ν, and k. However, keep in mind that only two of these are needed to characterize the mechanical behavior of material. Although we have developed a few relationships between various moduli, many other such relations can also be found. In fact, it can be shown that all five elastic constants are interrelated, and if any two are given, the remaining three can be determined by using simple formulas. Results of these relations are conveniently summarized in Table 4.5. This table should be marked for future reference, because it will prove to be useful for calculations throughout this textbook.

Table 4.5 Relations among different elastic constants

	E	ν	k	μ	λ
E, ν	E	ν	$\frac{E}{3(1-2\nu)}$	$\frac{E}{2(1+\nu)}$	$\frac{E}{(1+\nu)(1-2\nu)}$
E, k	E	$\frac{3k-E}{6k}$	k	$\frac{3kE}{9k-E}$	$\frac{3k(3k-E)}{9k-E}$
E, μ	E	$\frac{E-2\mu}{2\mu}$	$\frac{\mu E}{3(3\mu-E)}$	μ	$\frac{\mu(E-2\mu)}{3\mu-E}$
E, λ	E	$\frac{2\lambda}{E+\lambda+R}$	$\frac{E+3\lambda+R}{6}$	$\frac{E-3\lambda+R}{4}$	λ
ν, k	$3k(1 - 2\nu)$	ν	k	$\frac{3k(1-2\nu)}{2(1+\nu)}$	$\frac{3k\nu}{1+\nu}$
ν, μ	$2\mu(1 + \nu)$	ν	$\frac{2\mu(1+\nu)}{3(1-2\nu)}$	μ	$\frac{2\mu\nu}{1-2\nu}$
ν, λ	$\frac{\lambda(1+\nu)(1-2\nu)}{\nu}$	ν	$\frac{\lambda(1+\nu)}{3\nu}$	$\frac{\lambda(1-2\nu)}{2\nu}$	λ
k, μ	$\frac{9k\mu}{3k+\mu}$	$\frac{3k-2\mu}{6k+2\mu}$	k	μ	$k-\frac{2}{3}\mu$
k, λ	$\frac{9k(k-\lambda)}{3k-\lambda}$	$\frac{\lambda}{3k-\lambda}$	k	$\frac{3}{2}(k-\lambda)$	λ
μ, λ	$\frac{\mu(3\lambda+2\mu)}{\lambda+\mu}$	$\frac{\lambda}{2(\lambda+\mu)}$	$\frac{3\lambda+2\mu}{3}$	μ	λ

$R=\sqrt{E^2+9\lambda^2+2E\lambda}$

4.7.1 Simple Tension

Under the loading scenario of simple tension, the relationship among E, ν, λ and μ can be derived. Because $\sigma_y=\sigma_z=\tau_{yz}=\tau_{xz}=\tau_{xy}=0$, the strain components can be worked out using the strain-stress relationship as below:

$$\begin{cases} \varepsilon_x=\dfrac{\lambda+\mu}{\mu(3\lambda+2\mu)}\sigma_x \\ \varepsilon_y=\varepsilon_z=-\dfrac{\lambda}{2\mu(3\lambda+2\mu)}\sigma_x \\ \gamma_{yz}=\gamma_{xz}=\gamma_{xy}=0 \end{cases} \tag{4.82}$$

On the other hand, based on the definition of E and ν, i.e. their physical meanings, we can obtain

$$\begin{cases} \varepsilon_x=\dfrac{\sigma_x}{E} \\ \varepsilon_y=\varepsilon_z=-\dfrac{\nu}{E}\sigma_x \\ \gamma_{yz}=\gamma_{xz}=\gamma_{xy}=0 \end{cases} \tag{4.83}$$

Comparing Eq. (4.82) with Eq. (4.83), we have

$$E=\frac{\mu(3\lambda+2\mu)}{\lambda+\mu},\nu=\frac{\lambda}{2(\lambda+\mu)} \tag{4.84}$$

which can also be written as

$$\lambda = \frac{E\nu}{(1+\nu)(1-2\nu)}, \mu = \frac{E}{2(1+\nu)} \tag{4.85}$$

4.7.2 Pure Shear

Under the loading scenario of pure shear, the relationship between μ and G can be obtained. Because $\sigma_x = \sigma_y = \sigma_z = \tau_{yz} = \tau_{xz} = 0$, the strain components can be worked out using the strain-stress relationship as below:

$$\begin{cases} \varepsilon_x = \varepsilon_y = \varepsilon_z = \gamma_{yz} = \gamma_{xz} = 0 \\ \gamma_{xy} = \dfrac{\tau_{xy}}{\mu} \end{cases} \tag{4.86}$$

On the other hand, based on the physical meaning of G, we can have

$$\begin{cases} \varepsilon_x = \varepsilon_y = \varepsilon_z = \gamma_{yz} = \gamma_{xz} = 0 \\ \gamma_{xy} = \dfrac{\tau_{xy}}{G} \end{cases} \tag{4.87}$$

Comparing Eq. (4.86) with Eq. (4.87), we have

$$\mu = G \tag{4.88}$$

4.7.3 Hydrostatic Compression

Under the loading scenario of hydrostatic compression, the relationship between k, E and ν can be obtained. Because $\sigma_x = \sigma_y = \sigma_z = -p$, $\tau_{xy} = \tau_{yz} = \tau_{xz} = 0$, the strain components can be worked out using the strain-stress relationship as below:

$$\begin{cases} \varepsilon_x = \dfrac{1}{E}\left[\sigma_x - \nu\left(\sigma_y + \sigma_z\right)\right] \\ \varepsilon_y = \dfrac{1}{E}\left[\sigma_y - \nu\left(\sigma_x + \sigma_z\right)\right] \\ \varepsilon_z = \dfrac{1}{E}\left[\sigma_z - \nu\left(\sigma_x + \sigma_y\right)\right] \\ \varepsilon_{xy} = \varepsilon_{yz} = \varepsilon_{xz} = 0 \end{cases} \tag{4.89}$$

Then,

Table 4.6 Typical values of elastic constants for common engineering materials

	E (GPa)	ν	μ (GPa)	λ (GPa)	k (GPa)	α (10^{-6}/°C)
Aluminum	68.9	0.34	25.0	54.6	71.8	25.5
Concrete	27.6	0.20	11.5	7.7	15.3	11
Copper	89.6	0.34	33.4	71	93.3	18
Glass	68.9	0.25	27.6	27.6	45.9	8.8
Nylon	28.3	0.40	10.1	4.04	47.2	102
Rubber	0.0019	0.499	0.654×10^{-3}	0.326	0.326	200
Steel	207	0.29	80.2	111	164	13.5

$$(\varepsilon_x + \varepsilon_y + \varepsilon_z) = \frac{1}{E}\left[(1-2\nu)(\sigma_x + \sigma_y + \sigma_z)\right] \tag{4.90}$$

which can be written in the following format

$$\varepsilon_{kk} = \frac{1-2\nu}{E}\sigma_{kk} \tag{4.91}$$

On the other hand, based on the physical meaning of k, we have

$$\sigma_m = 3k\varepsilon_m \tag{4.92}$$

where,

$$\begin{cases} \sigma_m = \frac{1}{3}(\sigma_x + \sigma_y + \sigma_z) = \frac{1}{3}\sigma_{kk} \\ \varepsilon_m = \frac{1}{3}(\varepsilon_x + \varepsilon_y + \varepsilon_z) = \frac{1}{3}\varepsilon_{kk} \end{cases} \tag{4.93}$$

Comparing Eq. (4.91) with Eq. (4.92), we have

$$k = \frac{E}{3(1-2\nu)} \tag{4.94}$$

Typical nominal values of elastic constants for particular engineering materials are given in Table 4.6. These elastic constants represent average values, and some variations will occur for specific materials.

Chapter 5
Formulation and Solution Strategies

Yongtao Lyu and Yonggang Zheng

5.1 Introduction

The basic field equations in theory of elasticity have been developed in the previous chapters. These results comprise a system of differential and algebraic relations among the stresses, strains and displacements that express particular physics at all points within the body under investigation. In this chapter we now would like to complete the general formulation by first developing boundary conditions appropriate for use with the field equations. These conditions specify the physics that occur on the boundary of the body, and generally provide the loading inputs that physically create the interior stress, strain and displacement fields. Although the field equations are the same for all problems, boundary conditions are different for each problem. Therefore, proper development of boundary conditions is essential for problem solution, and thus it is important to acquire a good understanding of such development procedures. Combining field equations with boundary conditions then establishes the fundamental boundary value problems of the theory. This eventually leads us into two different formulations: one in terms of displacements and the other in terms of stresses. Because boundary value problems are difficult to solve, different strategies have been developed to solve the problems. In this chapter, we review in a general way several of these strategies.

Y. Lyu (✉) · Y. Zheng
School of Mechanics and Aerospace Engineering, Dalian University of Technology, Dalian, Liaoning, China
e-mail: Yongtaolu@dlut.edu.cn; zhengyg@dlut.edu.cn

Y. Lyu, Y. Zheng, *Theory of Elasticity*,
https://doi.org/10.1007/978-981-95-6043-1_5

5.2 Relevant Concepts

5.2.1 Type of Loads

There are three basic types of loads. The first is the point force or the concentrated force, which is the kind of force applied on one single point. The second is the distributed loads or pressure, which is applied on a line segment, an area or a volume. The third is the concentrated moment. Based on these three basic types of loads, there can be different combinations and then different mixed loads can be obtained. For example, a deformable body can be under the mixed loading of a point force and a body force (volumetric pressure).

5.2.2 Reaction Forces and Moments

When a deformable body is in the equilibrium state under external loading, the sum of all the forces on the object should be zero and the sum of all the moments should also be zero, i.e., all in balanced states. For example, in Fig. 5.1, an external point force is applied on the left end of the beam and the right end of the beam is fully constrained. To make the system balanced, the reaction forces and moments should be generated at the constrained side of the beam. The left end cannot be translated implying that the reaction forces can be generated at the support end. Furthermore, the left end cannot be rotated implying that reactions moments can be generated at the support end. From Fig. 5.1, it can be worked out that the total moments about the support point is:

$$M_T = M_R - FL$$

The sum of the moments should be zero, and thus the reaction moment can be worked out as:

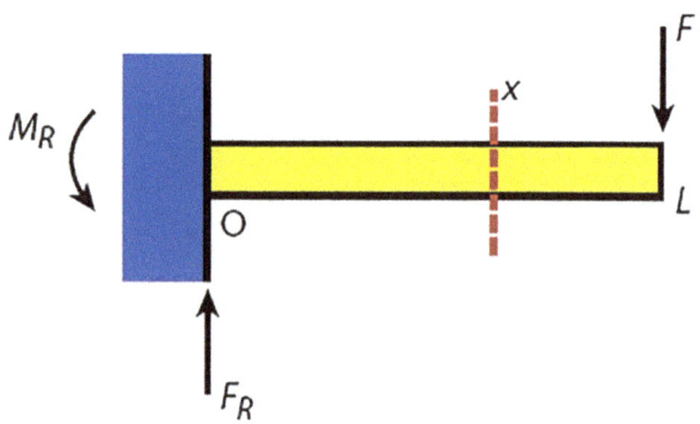

Fig. 5.1 Illustration of reaction forces and moments

$$M_R = FL$$

Furthermore, from the condition that all the forces in the system should be balanced, we can work out the reaction force as:

$$F_R = F$$

It should be noted that the reaction forces and moments can only be generated at the constrained parts of the deformable body. The forces and moments in other parts should either be the external loading or the internal forces and moments. For example, if a section is made at the red dotted line in Fig. 5.1, forces and moments at the cross-section should only be called the internal forces and internal moments.

5.2.3 Internal Forces and Moments

In contrast to the external forces and moments, the internal forces and moments are within the deformable body and all the points within the deformable body are associated with the internal forces and moments. The method to work out the internal forces and moments is to make a cut through at the position where the internal forces and moments need to be worked out. For example, to work out the internal forces and moments at the red dotted position in Fig. 5.2a, a cut through is made and two segments are obtained (Fig. 5.2b). Both segments are in balanced states, i.e., for the right segment, the reaction forces and moments are balanced with the internal forces and moments at the cross-section, and for the left segment, the external concentrated force is balanced with the internal forces and moments at the cross-section. From the

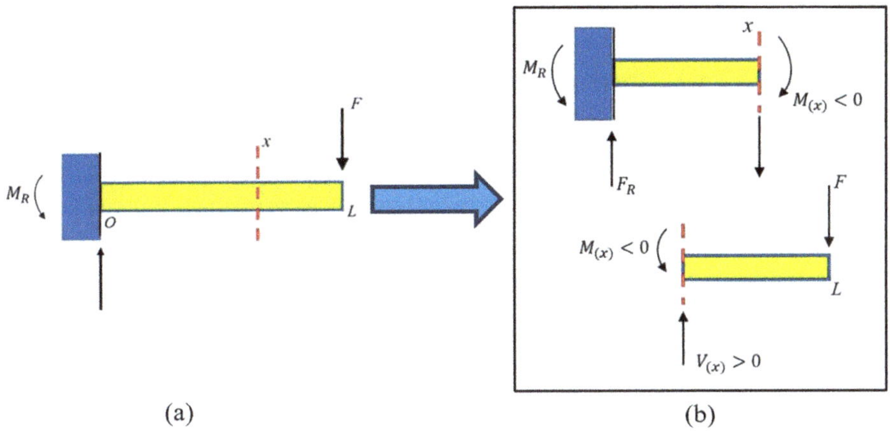

Fig. 5.2 Definition of the internal forces and moments: (**a**) Beam structure with external force and moment; (**b**) Sectioned beam structure

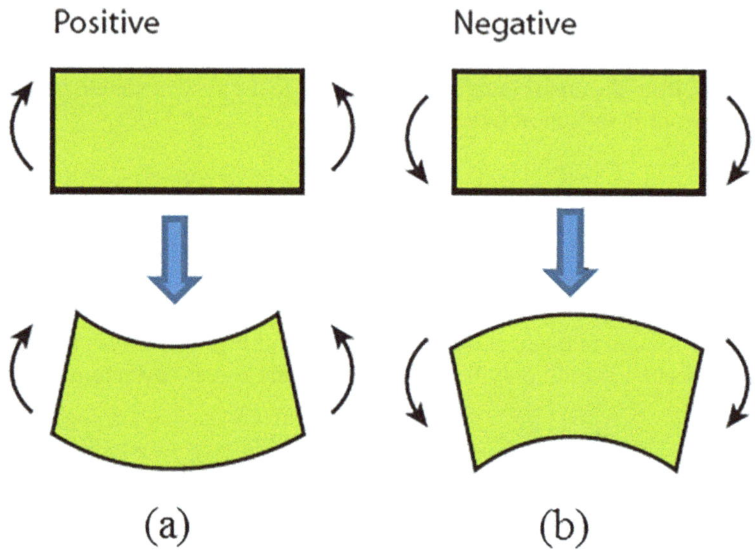

Fig. 5.3 Definition of the directions of the internal moments: (**a**) definition of positive internal moment; (**b**) definition of negative internal moment

equilibrium equations, the internal moment at the cross-section can be worked out as:

$$M(x) = -F(L - x)$$

The internal force at the cross-section is:

$$V(x) = F$$

Regarding the direction of the moments, if the moments are clockwise, they are positive (Fig. 5.3a). Otherwise (i.e., the moments are anti-clockwise), they are negative (Fig. 5.3b). Regarding the direction of the shear force, if the shear forces make the left side of the deformable segment upward and the right side downward, the shear forces are positive, i.e., the positive shear forces spin an element clockwise (Fig. 5.4a). Otherwise (i.e., the internal shear forces spin an element anti-clockwise), then they are negative (Fig. 5.4b)

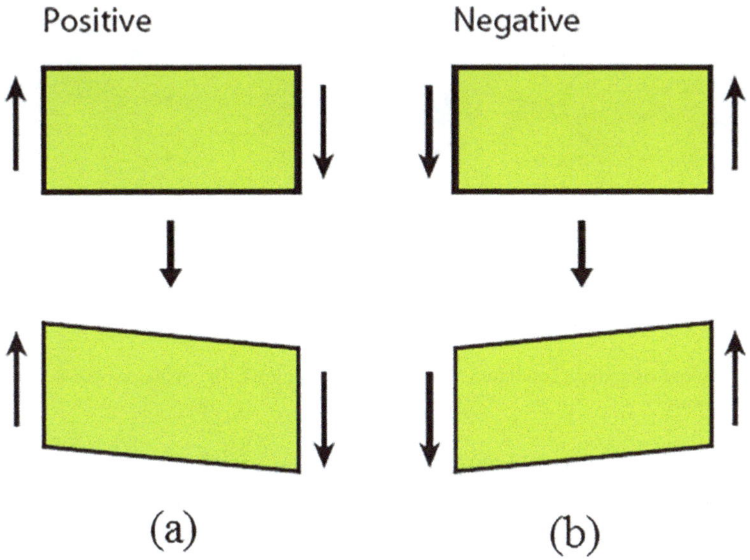

Fig. 5.4 Definition of the directions of the internal shear forces: (**a**) definition of positive shear force; (**b**) definition of negative shear force

5.3 Boundary Conditions and Problem Classifications

Before beginning the discussion on boundary conditions, we list here the basic field equations for linear isotropic elasticity.

Strain-displacement relations:

$$\varepsilon_{ij} = \frac{1}{2}\left(u_{i,j} + u_{j,i}\right) \tag{5.1}$$

Compatibility relations:

$$\varepsilon_{ij,kl} + \varepsilon_{kl,ij} - \varepsilon_{ik,jl} - \varepsilon_{jl,ik} = 0 \tag{5.2}$$

Equilibrium equations:

$$\sigma_{ij,j} + F_i = 0 \tag{5.3}$$

Elastic constitutive law (Hooke's law):

$$\begin{cases} \sigma_{ij} = \lambda \varepsilon_{kk} \delta_{ij} + 2\mu \varepsilon_{ij} \\ \varepsilon_{ij} = \dfrac{1+\nu}{E} \sigma_{ij} - \dfrac{\nu}{E} \sigma_{kk} \delta_{ij} \end{cases} \tag{5.4}$$

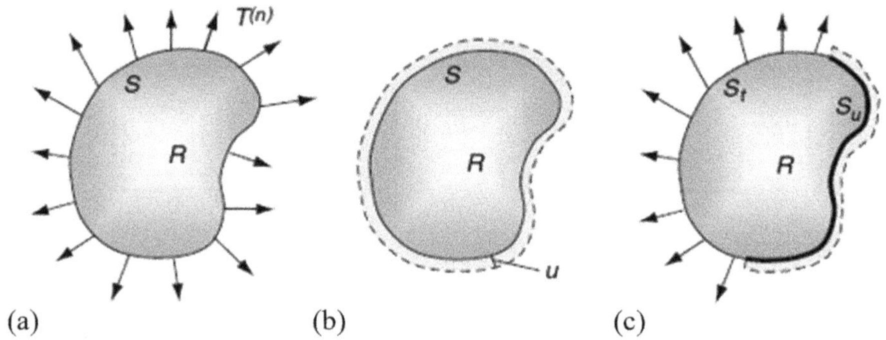

Fig. 5.5 Typical boundary conditions: (**a**) traction conditions; (**b**) displacement conditions; (**c**) mixed conditions

It has been mentioned previously that the compatibility relations ensure that the displacements are continuous and single-valued and are necessary only when the strains are arbitrarily specified. If, however, the displacements are included in the problem formulation, the solution normally generates single-valued displacements and the strain compatibility is automatically satisfied. Thus, in discussing the general system of equations of elasticity, the compatibility relations are normally put aside. Therefore, the general system of elasticity field equations refers to the fifteen relations presented in Eqs. (5.1), (5.3) and (5.4). It is convenient to define this entire system using a generalized operator notation as

$$\Im\{u_i, \varepsilon_{ij}, \sigma_{ij}; \lambda, \mu, F_i\} = 0 \tag{5.5}$$

This system involves fifteen unknowns including three displacements u_i, six strains ε_{ij} and six stresses σ_{ij}. The terms after the semicolon indicate that the system is also dependent on two elastic material constants (for isotropic materials) and on the density of the body force. It is reassuring that the number of equations matches the number of unknowns to be determined. However, this general system of equations is of such complexity that solutions by using analytical methods are essentially impossible and further simplification is required to solve problems of interest. Before proceeding with development of such simplifications, it is useful first to discuss typical boundary conditions connected with the elasticity model, and this leads us to the classification of the fundamental problems.

Similar to other field problems in engineering science (e.g., fluid mechanics, heat conduction, diffusion, electromagnetics), the solution of Eq. (5.5) requires appropriate boundary conditions on the body under investigation. The common types of boundary conditions for elasticity applications normally include specification of how the body is supported or loaded. This concept is mathematically formulated by specifying either the displacements or tractions at boundary points. Figure 5.5 illustrates this general idea for three typical cases including tractions, displacements, and a mixed case for which tractions are specified on boundary S_t and displacements are given on the remaining portion S_u such that the total boundary is given by $S = S_t + S_u$.

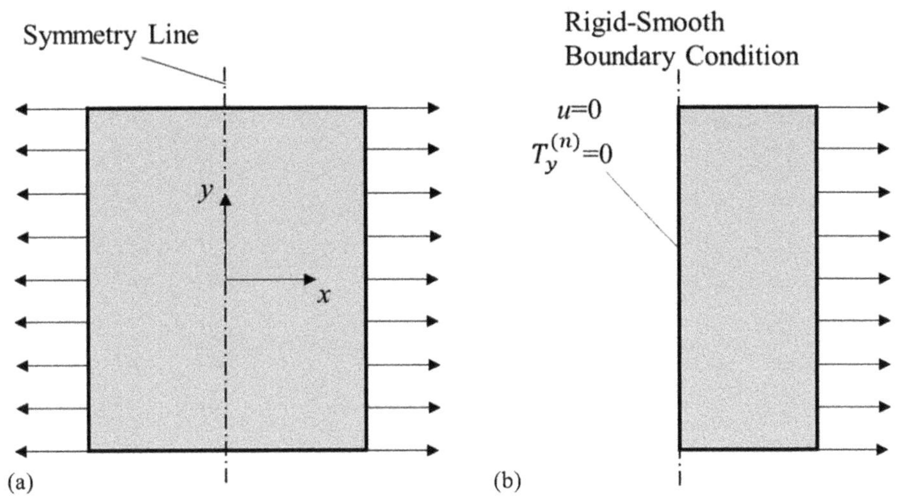

Fig. 5.6 Illustration of symmetric boundary condition: (**a**) Plane structure with a symmetry line; (**b**) Structure with symmetric boundary condition applied

Another type of mixed boundary condition can also occur. Although it is generally not possible to specify completely both the displacements and tractions at the same boundary point, it is possible to prescribe part of the displacement and part of the traction. Typically, this type of mixed condition involves the specification of a traction and displacement in two different orthogonal directions. A common example of this situation is shown in Fig. 5.6 for a case involving a surface of problem symmetry where the condition is one of a rigid-smooth boundary with zero normal displacement and zero tangential traction. Notice that in this example the body under investigation is subdivided along the symmetry line, thus creating a new boundary surface and resulting in a smaller region to be analyzed. Minimizing the size of the domain under study is commonly useful in computational modeling. However, determining expected symmetry properties in the solution can often be useful to simplify the analytical analyses as well.

Because boundary conditions play a very essential role in properly formulating and solving elasticity problems, it is important to acquire a clear understanding of their specification and usage. Improper specification results in either no solution or a solution to a different problem rather than the one which was originally sought. Boundary conditions are normally specified using the coordinate system describing the problem, and thus particular components of the displacements and tractions are set to be equal to prescribed values. For displacement conditions, such a specification is straightforward, and a common example includes fixed boundaries where the displacements are zeros. For traction boundary conditions, the specification can be a bit more complex.

Figure 5.7 illustrates particular cases in which the boundaries coincide with Cartesian or polar coordinate surfaces. The traction specification can be reduced to a stress specification. For the Cartesian example in which $y=$ constant, the normal traction becomes simply the stress component σ_y, while the tangential traction

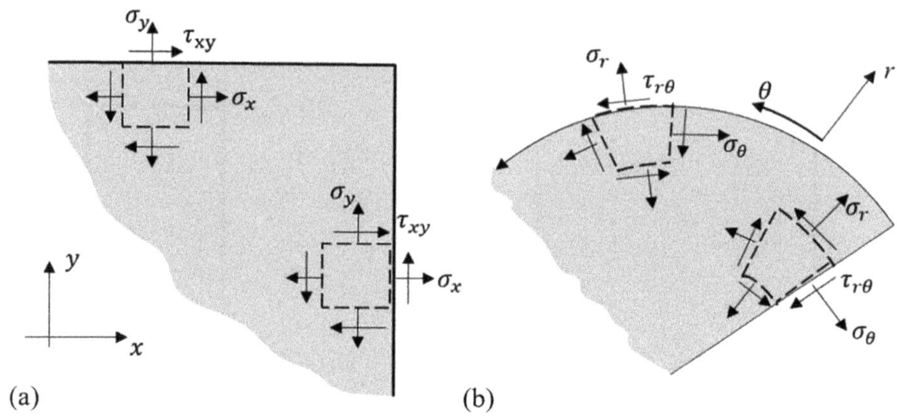

Fig. 5.7 Boundary stress components on coordinate surfaces: (**a**) Cartesian coordinate boundaries; (**b**) Polar coordinate boundaries

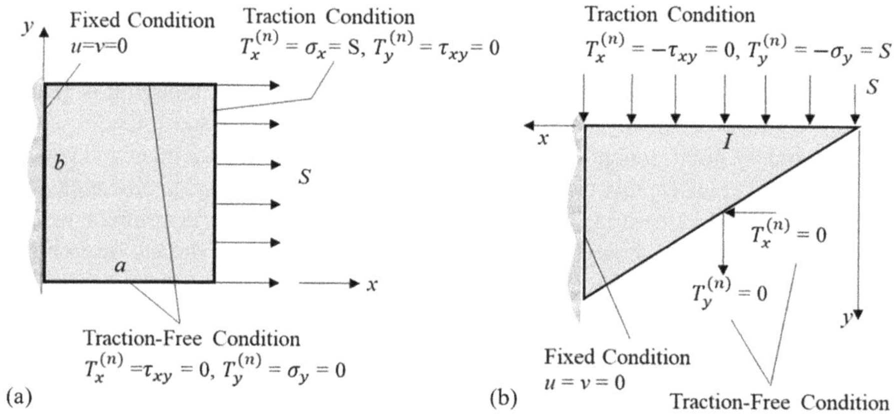

Fig. 5.8 Examples of boundary conditions: (**a**) coordinate surface boundaries; (**b**) non-coordinate surface boundaries

reduces to τ_{xy}. For this case, σ_x exists only inside the region; thus, this component of stress cannot be specified on the boundary surface $y=$ constant. A similar situation exists on the vertical boundary $x=$ constant, where the normal traction is now σ_x, the tangential traction is τ_{xy}, and the stress component σ_y exists inside the domain. Similar arguments can be made for polar coordinate boundary surfaces as shown. Drawing the appropriate element along the boundary as illustrated allows a clear visualization of the particular stress components that act on the surface in question. Such a sketch also allows determination of the positive directions of these boundary stresses, and this is useful to properly match with boundary loadings that might be prescribed. It is recommended that sketches similar to Fig. 5.7 be used to aid in the proper development of boundary conditions during problem formulation in later chapters.

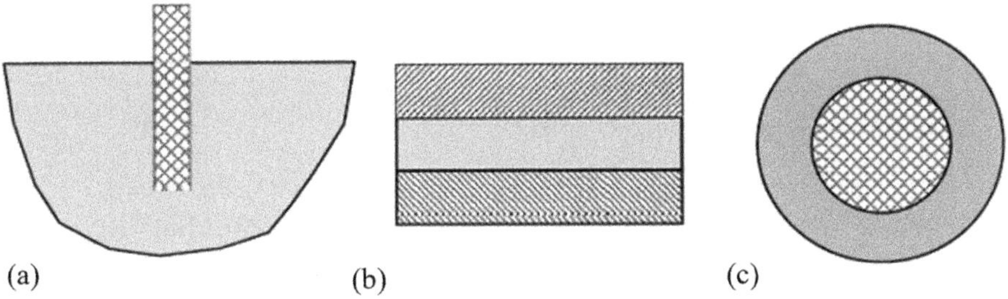

Fig. 5.9 Typical composite bodies: (**a**) embedded fiber; (**b**) layered composite plate; (**c**) composite cylinder

The pair of 2D example problems with mixed conditions as shown in Fig. 5.8 is considered. For the rectangular plate problem, all four boundaries are coordinate surfaces, and this simplifies specification of particular boundary conditions. The fixed conditions on the left edge simply require that x and y displacement components vanish on $x = 0$, and this specification does not change even if this were not a coordinate surface. However, as per our previous discussion, the traction conditions on the other three boundaries simplify because they are coordinate surfaces. These simplifications are shown in the figure for each of the traction-specified surfaces. The second problem of a tapered cantilever beam has an inclined face that is not a coordinate surface.

For this problem, the fixed end and top surface follow similar procedures as in the first example and are specified in the figure. However, on the inclined face, the traction is to be zero and this does not reduce to a simple specification of the vanishing of individual stress components. On this face each traction component is set to zero, giving the result

$$\left\{\begin{array}{l} T_x^{(n)} = \sigma_x n_x + \tau_{xy} n_y = 0 \\ T_y^{(n)} = \tau_{xy} n_x + \sigma_y n_y = 0 \end{array}\right. \tag{5.6}$$

where n_x and n_y are the components of the unit normal vector to the inclined face. This is the more general type of specification, and it should be clearly noted that none of the individual stress components in the xy system will vanish along this surface. It should also be pointed out for this problem that the components of unit normal vector are constants for all points on the inclined face. However, for curved boundaries, the normal vector changes with surface position. Note that using strain-displacement relations in Hooke's law allows the stresses and therefore tractions to be expressed in terms of displacement gradients. Thus, the traction boundary conditions can actually be expressed in terms of displacements if they are desired.

Another type of boundary condition formulation occurs for composite bodies composed of two or more pieces of different materials with different elastic moduli. Examples of such situations are common and include many types of composite materials and structures, as shown in Fig. 5.9. For such problems, the elasticity solution must be developed independently for each material body, thus requiring specification of external boundary conditions and internal interface conditions that

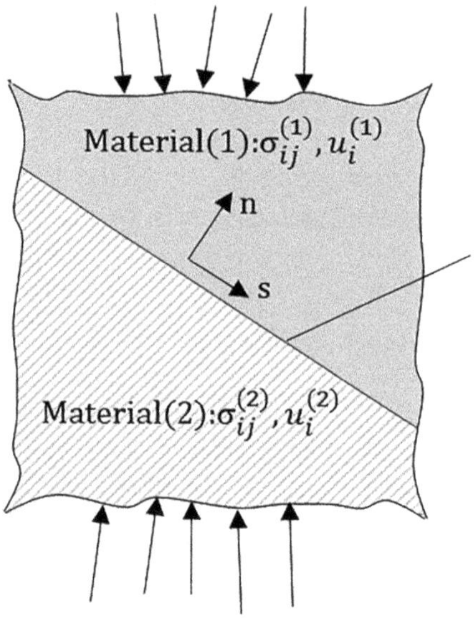

Fig. 5.10 Composite elastic continuum

exist between each material phase (Fig. 5.10). These interface conditions specify how the various composite pieces are joined together. Common simple examples include the perfectly bonded interface, where both the displacements and tractions are continuous at the interface. Another common case is the slip interface that prescribes continuity of normal components of displacement and traction, vanishing of the tangential traction, and allows a discontinuity in the tangential displacement. Other more complicated interface conditions have been proposed in the literature in an effort to model more sophisticated contact behaviors.

We are now in the position to formulate and classify the three fundamental boundary-value problems in the theory of elasticity that are related to solving the general system of field equations (Eq. (5.5)). Our presentation is limited to the static case.

Problem 1: Traction problem Determine the distribution of displacements, strains, and stresses in the interior of an elastic body in equilibrium when body forces are given and the distribution of tractions is prescribed over the surface of the body:

$$T_i^{(n)}\left(x_i^{(s)}\right) = f_i\left(x_i^{(s)}\right) \tag{5.7}$$

where $x_i^{(s)}$ denote boundary points and $f_i\left(x_i^{(s)}\right)$ are the prescribed traction values.

Problem 2: Displacement problem Determine the distribution of displacements, strains, and stresses in the interior of an elastic body in equilibrium when body forces are given and the distribution of displacements is prescribed over the surface of the body:

$$u_i\left(x_i^{(s)}\right) = g_i\left(x_i^{(s)}\right) \tag{5.8}$$

where $x_i^{(s)}$ denote boundary points and $g_i\left(x_i^{(s)}\right)$ are the prescribed displacement values.

Problem 3: Mixed problem Determine the distribution of displacements, strains, and stresses in the interior of an elastic body in equilibrium when body forces are given and the distribution of tractions is prescribed as per Eq. (5.7) over the surface S_t and the distribution of displacements is prescribed as per Eq. (5.8) over the surface S_u of the body.

As mentioned previously, the solution to any of these types of problems is difficult and further reduction and simplification of Eq. (5.5) are required for the development of the analytical solution method. Based on the description of Problem 1 with only traction boundary conditions, it would appear to be desirable to express the fundamental system solely in terms of stress, that is, $\mathfrak{J}^{(t)}\{\sigma_{ij}; \lambda, \mu, F_i\}$ thereby reducing the number of unknowns in the system. Likewise for Problem 2, a displacement-only formulation of the form $\mathfrak{J}^{(u)}\{u_i; \lambda, \mu, F_i\}$ would appear to simplify the problem. We now pursue such specialized formulations and explicitly determine these reduced field equation systems.

5.3.1 *Stress Formulation*

For the first fundamental problem in elasticity, the boundary conditions are to be given only in terms of the tractions or stress components. In order to develop solution methods for this case, it is very helpful to reformulate the general system presented in Eq. (5.5) by eliminating the displacements and strains and thereby cast a new system solely in terms of the stresses. We now develop this reformulated system. By eliminating the displacements, we must now include the compatibility equations in the fundamental system of field equations. We therefore start by using Hooke's law and eliminate the strains in the compatibility relations to get

$$\begin{aligned}\sigma_{ij,kk} + \sigma_{kk,ij} - \sigma_{ik,jk} - \sigma_{jk,ik} &= \frac{\nu}{1+\nu} \\ &\times \left(\sigma_{mm,kk}\delta_{ij} + \sigma_{mm,ij}\delta_{kk} - \sigma_{mm,jk}\delta_{ik} - \sigma_{mm,ik}\delta_{jk}\right)\end{aligned} \tag{5.9}$$

Although Eq. (5.9) represents the compatibility in terms of stress, a more useful result is found by incorporating the equilibrium equations into the system. Recall that from Eq. (5.3), $\sigma_{ij,j} = -F_i$, and also note that $\delta_{kk} = 3$. Substituting these results into Eq. (5.9) gives

$$\sigma_{ij,kk} + \frac{1}{1+\nu}\sigma_{kk,ij} = \frac{\nu}{1+\nu}\sigma_{mm,kk}\delta_{ij} - F_{i,j} - F_{j,i} \tag{5.10}$$

For the case $i = j$, Eq. (5.10) reduces to $\sigma_{ii,kk} = -\frac{1+\nu}{1-\nu}F_{i,i}$. Substituting this result back into Eq. (5.10) gives the desired relation

$$\sigma_{ij,kk} + \frac{1}{1+\nu}\sigma_{kk,ij} = -\frac{\nu}{1-\nu}\delta_{ij}F_{k,k} - F_{i,j} - F_{j,i} \tag{5.11}$$

This result is the compatibility relations in terms of the stress and is commonly called the *Beltrami-Michell compatibility equations*. For the case with no body forces, these relations can be expressed as the following six scalar equations:

$$, \begin{cases} (1+\nu)\nabla^2\sigma_x + \frac{\partial^2}{\partial x^2}(\sigma_x + \sigma_y + \sigma_z) = 0 \\ (1+\nu)\nabla^2\sigma_y + \frac{\partial^2}{\partial y^2}(\sigma_x + \sigma_y + \sigma_z) = 0 \\ (1+\nu)\nabla^2\sigma_z + \frac{\partial^2}{\partial z^2}(\sigma_x + \sigma_y + \sigma_z) = 0 \\ (1+\nu)\nabla^2\tau_{xy} + \frac{\partial^2}{\partial x \partial y}(\sigma_x + \sigma_y + \sigma_z) = 0 \\ (1+\nu)\nabla^2\tau_{yz} + \frac{\partial^2}{\partial y \partial z}(\sigma_x + \sigma_y + \sigma_z) = 0 \\ (1+\nu)\nabla^2\tau_{zx} + \frac{\partial^2}{\partial z \partial x}(\sigma_x + \sigma_y + \sigma_z) = 0 \end{cases} \tag{5.12}$$

Combining these results with the three equilibrium equations (Eq. (5.3)) provides the necessary six relations to solve for the six unknown stress components for the general 3D case. This system constitutes the stress formulation for elasticity theory and is appropriate for use with traction boundary condition problems. Once the stresses have been determined, the strains may be found from the Hooke's law, and the displacements can then be computed through the integration of Eq. (5.1). It should be noted that, such an integration process determines the displacements only up to an arbitrary rigid-body motion, and the displacements obtained are single-valued only if the region under study is simply connected.

The system of equations for the stress formulation is still rather complex, and analytical solutions are commonly determined for this case by making use of stress functions. This concept establishes a representation for the stresses that automatically satisfies the equilibrium equations. For the 2D case, this concept represents the in-plane stresses in terms of a single function. The representation satisfies equilibrium, and the remaining compatibility equations yield a single partial differential equation in terms of the stress function. Having reduced the system to a single equation, we can employ many analytical methods to find solutions of interest.

5.3.2 Displacement Formulation

We now would like to develop the reduced set of field equations solely in terms of the displacements. This system is referred to as the displacement formulation and is most useful when combined with displacement-only boundary conditions. This development is somewhat more straightforward than our previous discussion for the stress formulation. For this case, we would like to eliminate the strains and stresses from the fundamental system. This is easily accomplished by using the strain-displacement relations in Hooke's law to give

$$\sigma_{ij} = \lambda u_{k,k}\delta_{ij} + \mu\left(u_{i,j} + u_{j,i}\right) \tag{5.13}$$

which can be expressed as six scalar equations

$$\left\{\begin{aligned} \sigma_x &= \lambda\left(\frac{\partial u}{\partial x} + \frac{\partial v}{\partial y} + \frac{\partial w}{\partial z}\right) + 2\mu\frac{\partial u}{\partial x} \\ \sigma_y &= \lambda\left(\frac{\partial u}{\partial x} + \frac{\partial v}{\partial y} + \frac{\partial w}{\partial z}\right) + 2\mu\frac{\partial v}{\partial y} \\ \sigma_z &= \lambda\left(\frac{\partial u}{\partial x} + \frac{\partial v}{\partial y} + \frac{\partial w}{\partial z}\right) + 2\mu\frac{\partial w}{\partial z} \\ \tau_{xy} = \mu\left(\frac{\partial u}{\partial y} + \frac{\partial v}{\partial x}\right)&, \tau_{yz} = \mu\left(\frac{\partial v}{\partial z} + \frac{\partial w}{\partial y}\right), \tau_{zx} = \mu\left(\frac{\partial w}{\partial x} + \frac{\partial u}{\partial z}\right) \end{aligned}\right. \tag{5.14}$$

Using these relations in the equilibrium equations gives the result

$$\mu u_{i,k} + (\lambda + \mu)u_{k,k} + F_i = 0 \tag{5.15}$$

which are the equilibrium equations in terms of the displacements and is referred to as Navier's or Lame's equations. This system can be expressed in vector format as

$$\mu\nabla^2\boldsymbol{u} + (\lambda + \mu)\nabla(\nabla\cdot\boldsymbol{u}) + \boldsymbol{F} = 0 \tag{5.16}$$

or written out in terms of three scalar equations

$$\left\{\begin{aligned} &\mu\nabla^2 u + (\lambda + \mu)\frac{\partial}{\partial x}\left(\frac{\partial u}{\partial x} + \frac{\partial v}{\partial y} + \frac{\partial w}{\partial z}\right) + F_x = 0 \\ &\mu\nabla^2 v + (\lambda + \mu)\frac{\partial}{\partial y}\left(\frac{\partial u}{\partial x} + \frac{\partial v}{\partial y} + \frac{\partial w}{\partial z}\right) + F_y = 0 \\ &\mu\nabla^2 w + (\lambda + \mu)\frac{\partial}{\partial z}\left(\frac{\partial u}{\partial x} + \frac{\partial v}{\partial y} + \frac{\partial w}{\partial z}\right) + F_z = 0 \end{aligned}\right. \tag{5.17}$$

where the Laplacian is given by $\nabla^2 = (\partial^2/\partial x^2) + (\partial^2/\partial y^2) + (\partial^2/\partial z^2)$. It should be noted that for the case with no body forces, Navier's equations are expressible in terms of a single elastic constant (Poisson's ratio).

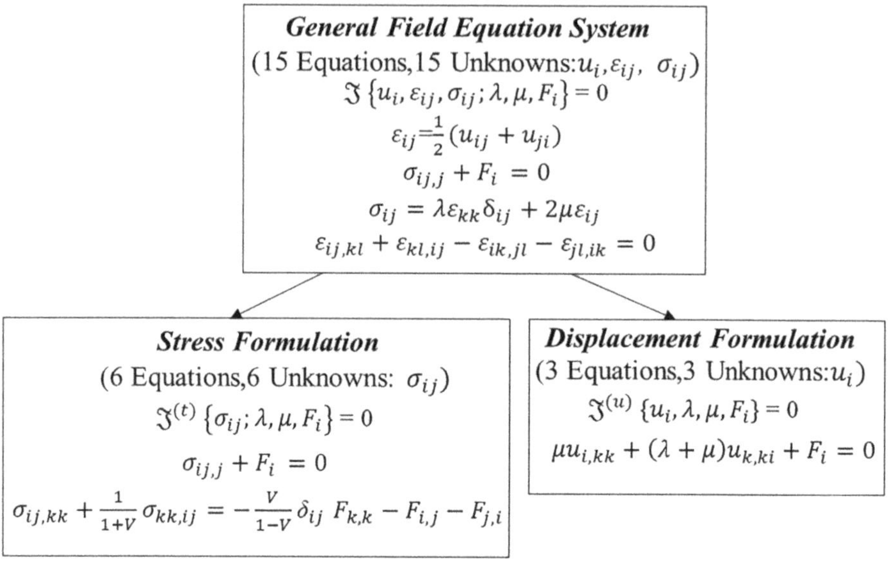

Fig. 5.11 Schematic representation of elasticity field equations

Navier's equations are the desired formulation for the displacement problem, and the system represents three equations for the three unknown displacement components. Similar to the stress formulation, this system is still difficult to solve, and additional mathematical techniques are developed to further simplify these equations for problem solution. Common methods normally employ the use of displacement potential functions. It should be noted that several such schemes can be developed that allow the displacement vector to be expressed in terms of particular potentials. These schemes generally simplify the problem by yielding uncoupled governing equations in terms of the displacement potentials. This then allows several analytical methods to be employed to solve problems of interest.

To help acquire a general understanding of these results, a flowchart summarizing the developed stress and displacement formulations, is shown in Fig. 5.11. Note that for the stress formulation, the resulting system $\Im^{(t)}\{\sigma_{ij}; \lambda, \mu, F_i\}$ is actually dependent on only the single material constant, i.e., Poisson's ratio, and thus it could be expressed as $\Im^{(t)}\{\sigma_{ij}; \nu, F_i\}$.

5.3.3 *Principle of Superposition*

A very useful tool for the solution to many problems in engineering science is the principle of superposition. This technique applies to any problem that is governed by linear equations. Under the assumption of small deformations and linear elastic constitutive behavior, all elasticity field equations are linear. Furthermore, the

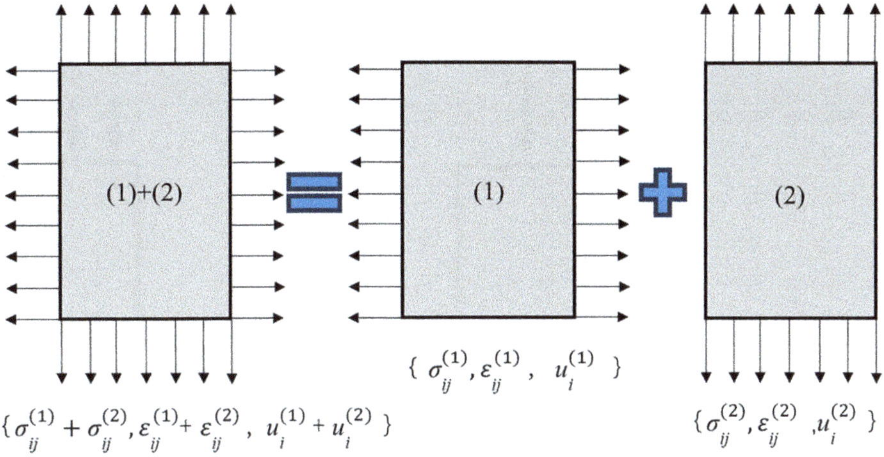

Fig. 5.12 Example of two-dimensional superposition

usual boundary conditions are also linear. Thus, under these conditions all governing equations are linear, and the superposition concept can be applied. The general statement of the principle can be expressed as follows:

For a given problem domain, if the state $\left\{\sigma_{ij}^{(1)}, \varepsilon_{ij}^{(1)}, u_i^{(1)}\right\}$ is a solution to the fundamental elasticity equations with prescribed body forces $F_i^{(1)}$ and surface tractions $T_i^{(1)}$, and the state $\left\{\sigma_{ij}^{(2)}, \varepsilon_{ij}^{(2)}, u_i^{(2)}\right\}$ is a solution to the fundamental equations with prescribed body forces $F_i^{(2)}$ and surface tractions $T_i^{(2)}$, then the state $\left\{\sigma_{ij}^{(1)} + \sigma_{ij}^{(2)}, \varepsilon_{ij}^{(1)} + \varepsilon_{ij}^{(2)}, u_i^{(1)} + u_i^{(2)}\right\}$ will be a solution to the problem with body forces $F_i^{(1)} + F_i^{(2)}$ and surface tractions $T_i^{(1)} + T_i^{(2)}$.

To see a more direct application of this principle, a simple 2D case with no body forces as shown in Fig. 5.12 is considered. It can be observed that the solution to the more complicated biaxial loading case (1) + (2) is thus found by adding the two simpler problems. This is a common application of the superposition principle, and we make repeated use of it throughout this textbook. Thus, once the solutions to some simple problems are generated, we can combine these results to generate a solution to a more complicated case with similar geometry.

5.3.4 Saint-Venant's Principle

Considering the set of three identical rectangular strips under compressive loadings shown in Fig. 5.13, as indicated, the only difference between each problem is the loading. Because the total resultant load applied to each problem is identical

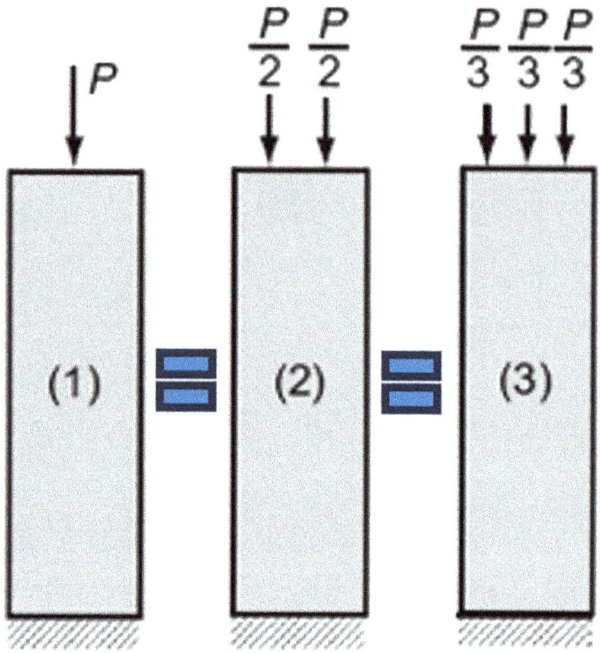

Fig. 5.13 Statically equivalent loadings

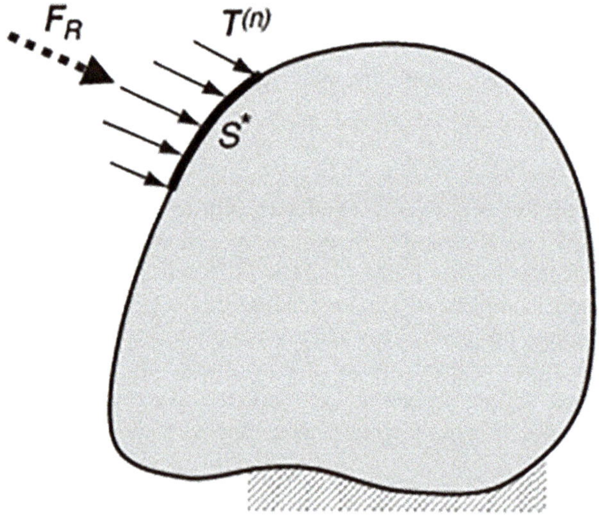

Fig. 5.14 Illustration of Saint-Venant's principle

(statically equivalent loadings), it is expected that the resulting stress, strain, and displacement fields near the bottom of each strip would be approximately the same.

This behavior can be generalized by considering an elastic solid with an arbitrary loading $T^{(n)}$ over a boundary portion S^*, as shown in Fig. 5.14. Based on experience from other field problems in engineering science, it seems logical that the particular boundary loading would produce detailed and characteristic effects only in the vicinity of S^*. In other words, we expect that at points far away from S^* the stresses generally depend more on the resultant F_R of the tractions rather than on the exact distribution. Thus, the characteristic signature of the generated stress, strain, and displacement fields from a given boundary loading tend to disappear as we move away from the boundary loading points. These concepts form the Saint-Venant's principle, which can be stated as follows:

The stress, strain, and displacement fields caused by two different statically equivalent force distributions on parts of the body far away from the loading points are approximately the same.

This statement of the principle includes qualitative terms such as far away and approximately the same, and thus does not provide quantitative estimates of the differences between the two elastic fields in question.

If we restrict our solution to points away from the boundary loading, Saint-Venant's principle allows us to change the given boundary conditions to a simpler statically equivalent statement and not affect the resulting solution. Such a simplification of boundary conditions greatly increases our chances of finding an analytical solution to the problem. This concept therefore proves to be very useful.

5.4 General Solution Strategies

Having completed our formulation and related solution principles, we now want to examine some general solution strategies commonly used to solve elasticity problems. At this stage, we categorize particular methods and outline only typical techniques that are commonly used. We first distinguish three general methods of solution called direct, inverse, and semi-inverse methods. Then we briefly discuss analytical, approximate, and numerical solution procedures.

5.4.1 Direct Method

This method seeks to determine the solution by direct integration of the field equations (Eq. (5.5)) or equivalently the stress and/or displacement formulations. Boundary conditions are to be satisfied exactly. This method normally encounters significant mathematical difficulties, thus limiting its application to problems with simple geometry.

Fig. 5.15 Prismatic bar under self-weight

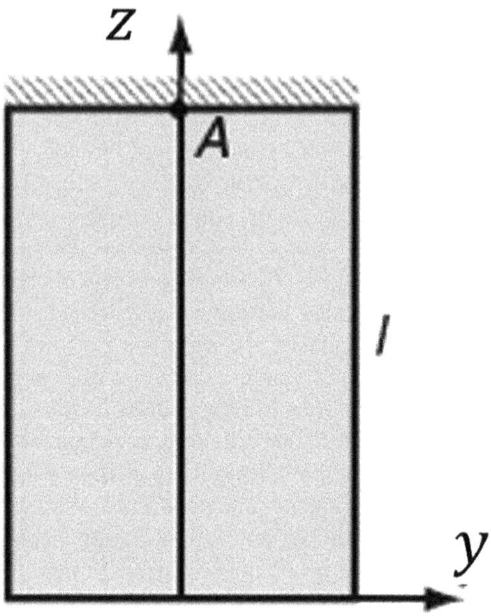

Example 5.1: Stretching Analysis of Prismatic Bar Under Its Own Weight Using the Direct Method

Question: As an example of a simple direct integration problem, the case of a uniform prismatic bar stretched by its own weight, as shown in Fig. 5.15, is considered. The body forces for this problem are $F_x = F_y = 0$, $F_z = -\rho g$, where ρ is the material mass density and g is the acceleration of gravity.

Solution: Assuming that on each cross-section we have uniform tension produced by the weight of the lower portion of the bar, the stress field would take the form

$$\sigma_x = \sigma_y = \tau_{xy} = \tau_{yz} = \tau_{zx} = 0$$

The equilibrium equations reduce to the simple result

$$\frac{\partial \sigma_z}{\partial z} = -F_z = \rho g$$

This equation can be integrated directly, and applying the boundary condition $\sigma_z = 0$ at $z = 0$ gives the result $\sigma_z(z) = \rho g z$. Next, using Hooke's law, the strains are easily calculated as

$$\varepsilon_z = \frac{\rho g z}{E}, \varepsilon_x = \varepsilon_y = -\frac{\nu \rho g z}{E}$$
$$\varepsilon_{xy} = \varepsilon_{yz} = \varepsilon_{xz} = 0$$

The displacements follow from integrating the strain-displacement relations, and for the case with boundary conditions of zero displacement and rotation at point A ($x = y = 0, z = l$), the final result is

$$u = -\frac{\nu \rho g x z}{E}, v = -\frac{\nu \rho g y z}{E}$$
$$w = \frac{\rho g}{2E}\left[z^2 + \nu\left(x^2 + y^2\right) - l^2\right]$$

5.4.2 *Inverse Method*

For this technique, particular displacements or stresses are selected that satisfy the basic field equations. A search is then conducted to identify a specific problem that would be solved by this solution field. This amounts to determining appropriate problem geometry, boundary conditions, and body forces that would enable the solution to satisfy all conditions on the problem. Using this scheme, it is sometimes difficult to construct solutions to a specific problem of practical interest.

5.4.3 *Semi-inverse Method*

In this scheme, part of the displacement and/or stress field is specified, and the other remaining portion is determined by the fundamental field equations (normally using direct integration) and the boundary conditions. It is often the case that constructing appropriate displacement and/or stress solution fields can be guided by approximate strength of materials theory. The usefulness of this approach is greatly enhanced by employing Saint-Venant's principle, whereby a complicated boundary condition can be replaced by a simpler statically equivalent distribution.

Example 5.2: Torsion Analysis of Prismatic Bars Using the Semi-inverse Method
Question: The following displacement field is proposed:

$$u = -\alpha y z, v = \alpha x z, w = w(x, y)$$

where α is a constant. The assumed field specifies the x and y components of displacement, while the z component is left to be determined as a function of the indicated spatial variables.

Solution: By using the strain-displacement relations and Hooke's law, the stress field corresponding to the given displacement field is given by

$$\begin{cases} \sigma_x = \sigma_y = \sigma_z = \tau_{xy} = 0 \\ \tau_{xz} = \mu\left(\dfrac{\partial w}{\partial x} - \alpha y\right) \\ \tau_{yz} = \mu\left(\dfrac{\partial w}{\partial y} + \alpha x\right) \end{cases}$$

Using these stresses into the equations of equilibrium gives the following result:

$$\frac{\partial^2 w}{\partial x^2} + \frac{\partial^2 w}{\partial y^2} = 0$$

which is actually the form of Navier's equations for this case. This result represents a single equation (Laplace's equation) to determine the unknown part of the assumed solution form. It should be apparent that by assuming part of the solution field, the remaining equations to be solved are greatly simplified. A specific domain in the $x - y$ plane along with appropriate boundary conditions is needed to complete the solution to a particular problem. Once this has been accomplished, the assumed solution form given in this example has been shown to satisfy all the field equations of elasticity.

There are numerous mathematical techniques used to solve the elasticity field equations. Many techniques involve the development of exact analytical solutions, while others involve the construction of approximate solution schemes. A third procedure involves the establishment of numerical solution methods.

5.4.4 Analytical Solution Procedures

A variety of analytical solution methods are used to solve the elasticity field equations. The following sections outline some of the more common methods.

Power series method: For many 2D elasticity problems, the stress formulation leads to the use of a stress function $\phi(x, y)$. It is shown that the entire set of field equations reduces to a single partial differential equation in tems of this stress function. A general mathematical scheme to solve this equation is to look for solutions in terms of a power series in the independent variables, that is, $\phi(x, y) = \sum C_{mn} x^m y^n$. The boundary conditions are used to determine the coefficients and number of terms to be used in the series.

Fourier method: A general scheme to solve a large variety of elasticity problems employs the Fourier method. This procedure is normally applied to the governing partial differential equations by using separation of variables, superposition, and Fourier series or Fourier integral theory. Although this is an ad hoc method, it has been shown to provide solutions to a large class of problems.

Integral transform method: A very useful mathematical technique to solve partial differential equations is the use of integral transforms. By applying a particular linear integral transformation to the governing equations, certain differential forms can be simplified or eliminated, thus allowing simple solution for the unknown transformed variables. Through appropriate inverse transformation, the original unknowns are retrieved, giving the required solution. Typical transforms that have been successfully applied to elasticity problems include Laplace, Fourier, and Hankel transforms.

Complex variable method: Several classes of problems in elasticity can be formulated in terms of functions of a complex variable. These include 2D plane problems, the torsion problem, and some particular thermoelastic cases. The complex variable formulation is very powerful and useful because many solutions can be found that would be intractable by other techniques.

5.4.5 Approximate Solution Procedures

With the recognized difficulty in finding exact analytical solutions, considerable work has been done to develop approximate solutions to elasticity problems. Much of this work has been in the area of variational methods, which are related to energy theorems. The principal idea of this approach is the connection of the elasticity field equations with a variational problem of finding an extremum of a particular integral functional. One specific technique is outlined below.

Ritz method: This scheme employs a set of approximating functions to solve elasticity problems by determining stationary values of a particular energy integral. The set of approximating functions is chosen to satisfy the boundary conditions on the problem, but only approximately make the energy integral take on an extremum. By including more terms in the approximating solution set, accuracy of the scheme is improved. Because of the difficulty in finding proper approximating functions for problems of complex geometry, variational techniques have made only limited contributions to the solution of general problems. However, they have made very important applications in the finite element method.

5.4.6 Numerical Solution Procedures

Over the past several decades, numerical methods have played a primary role in developing solutions to elasticity problems of complex geometry. Various schemes have been theoretically developed, and numerous private and commercial codes have been written for implementation on a variety of computer platforms. Several of the important methods are briefly discussed below.

Finite difference method: The finite difference method (FDM) replaces derivatives in the governing field equations by difference quotients, which involve values of the solution at discrete mesh points in the domain under investigation. Repeated applications of this representation set up algebraic systems of equations in terms of unknown mesh point values. The method is a classical one, having been established almost a century ago. The major difficulty with this scheme lies in the inaccuracies in dealing with regions of complex shape, although this problem can be eliminated through the use of coordinate transformation techniques.

Finite element method: The fundamental concept of the finite element method (FEM) lies in dividing the body under study into a finite number of pieces (subdomains) called elements. Particular assumptions are then made on the variation of the unknown dependent variable(s) across each element using so called interpolation or shape functions. This approximated variation is quantified in terms of solution values at special locations within the element called nodes. Through this discretization process, the method sets up an algebraic system of equations for unknown nodal values that approximates the continuous solution. Because element size and shape are variable, the finite element method can handle problem domains of complicated shape, and thus it has become a very useful and practical tool.

Boundary element method: The boundary element method (BEM) is based upon an integral statement of the governing equations of elasticity. The integral statement may be cast into a form that contains unknowns only over the boundary of the body domain. This boundary integral equation may then be solved by using concepts from the finite element method; that is, the boundary may be discretized into a number of elements and the interpolation approximation concept can then be applied. This method again produces an algebraic system of equations to solve for the unknown boundary nodal values, and the system is generally much smaller than that generated by internal discretization such as the finite element method. By avoiding interior discretization, the boundary element method has significant advantages over finite element schemes for infinite or very large domains and for cases in which only boundary information is sought.

5.5 Strain Energy and Related Principles

After the introduction of stress, strain and boundary conditions, below the associated concepts of work and energy are explored. Boundary tractions and body forces will do work on an elastic solid and this work will be stored inside the material in the form of strain energy. For the elastic case, removal of these loadings results in complete recovery of the stored energy within the body. Development of strain energy concepts can yield new and useful information not found by other methods. This study also leads to some new energy methods or principles that provide additional techniques to solve elasticity problems. In some sense these methods may be thought of as replacements of particular field equations that have been previously derived. For problems in structural mechanics involving rods, beams, plates and shells, energy methods have proved to be very useful in developing the governing equations and associated boundary conditions. These schemes have also provided a method to generate approximate solutions to elasticity problems. More recently, particular energy and variational techniques have been used extensively in the development of finite and boundary element analysis.

5.5.1 Strain Energy

As mentioned, the work done by surface and body forces on an elastic solid are stored inside the body in the form of strain energy. For an idealized elastic body, this stored energy is completely recoverable when the solid is returned to its original unstrained configuration. In order to quantify this behavior, we now would like to determine the strain energy in terms of the resulting stress and strain fields within the elastic solid. Considering first the simple uniform uniaxial deformation case with no body forces, as shown in Fig. 5.16, the cubical element of dimensions, i.e., dx, dy, dz, is under the action of a uniform normal stress in the x direction.

During this deformation process, we assume that the stress increases slowly from zero to six, such that inertia effects can be neglected. The strain energy stored is equal to the net work done on the element, and this is given by the following equation.

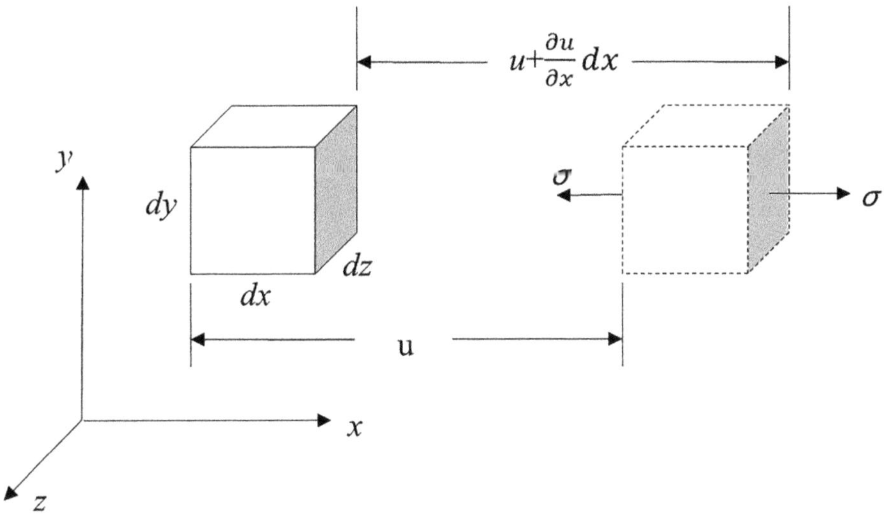

Fig. 5.16 Deformation under uniform uniaxial stress

$$dU = \int_0^{\sigma_x} \sigma d\left(u + \frac{\partial u}{\partial x}dx\right)dydz - \int_0^{\sigma_x} \sigma dudydz = \int_0^{\sigma_x} \sigma d\left(\frac{\partial u}{\partial x}\right)dxdydz \quad (5.18)$$

Using the strain displacement relation and Hooke's law,

$$\frac{\partial u}{\partial x} = \varepsilon_x = \frac{\sigma_x}{E}$$

and thus Eq. (5.18) can be reduced to

$$dU = \int_0^{\sigma_x} \sigma \frac{d\sigma}{E} dxdydz = \frac{\sigma_x^2}{2E} dxdydz \quad (5.19)$$

The strain energy per unit volume, or strain energy density, is specified by

$$U = \frac{dU}{dxdydz} = \frac{\sigma_x^2}{2E} = \frac{E\varepsilon_x^2}{2} = \frac{1}{2}\sigma_x\varepsilon_x \quad (5.20)$$

This result can be interpreted from the stress-strain curve shown in Fig. 5.17. Because the material is linear elastic, the strain energy for the uniaxial case is simply the shaded area under the stress-strain curve.

We next investigate the strain energy caused by the uniform shear stress. Choosing the same cubical element as previously analyzed and considering the case under uniform τ_{xy} and τ_{yx} loading, as shown in Fig. 5.18, following similar analyses, the strain energy is found to be

$$dU = \frac{1}{2}\tau_{xy}dydz\left(\frac{\partial v}{\partial x}dx\right) + \frac{1}{2}\tau_{yx}dxdz\left(\frac{\partial u}{\partial y}dy\right) = \frac{1}{2}\tau_{xy}\left(\frac{\partial u}{\partial y} + \frac{\partial v}{\partial x}\right)dxdydz \quad (5.21)$$

and thus, the strain energy density can be expressed as

Fig. 5.17 Strain energy for uniaxial deformation

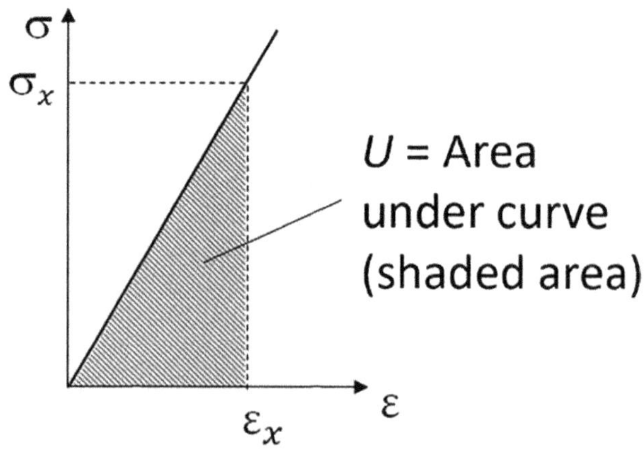

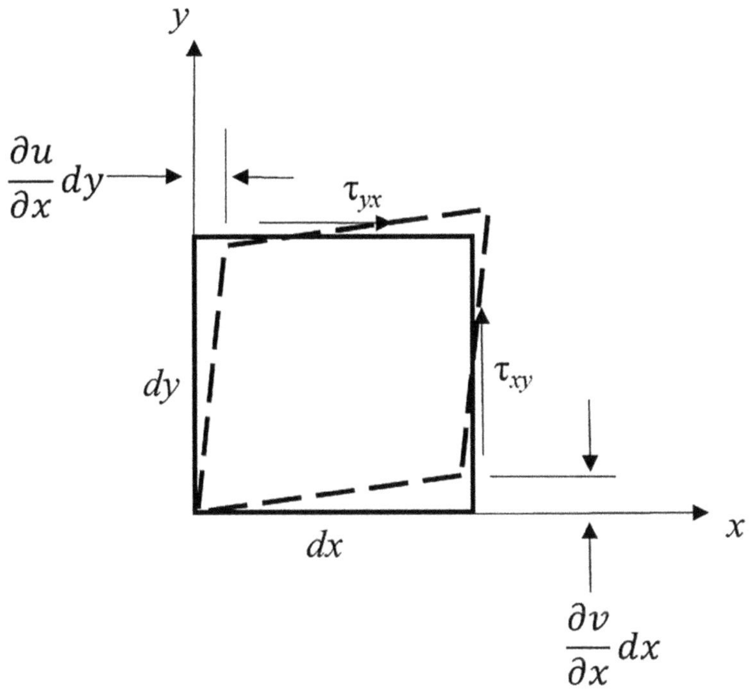

Fig. 5.18 Deformation under simple shear deformation

$$U = \frac{1}{2}\tau_{xy}\gamma_{xy} = \frac{\tau_{xy}^2}{2\mu} = \frac{\mu\gamma_{xy}^2}{2} \tag{5.22}$$

Equations (5.20) and (5.22) indicate that the strain energy is not a linear function of the stresses or strains, and thus the principle of superposition cannot be directly applied to develop the strain energy for a multidimensional state of stress. However, from conservation of energy, the work done does not depend on the order of loading application, but only on the final magnitudes of the stresses and strains. This concept then allows the normal and shear loadings to be applied one at a time and produces an additive total strain energy for a general 3D state of stress and strain, as follows:

$$U = \frac{1}{2}\sigma_{ij}e_{ij} = \frac{1}{2}\left(\sigma_x\varepsilon_x + \sigma_y\varepsilon_y + \sigma_z\varepsilon_z + \tau_{xy}\gamma_{xy} + \tau_{yz}\gamma_{yz} + \tau_{zx}\gamma_{zx}\right) \tag{5.23}$$

Although the preceding results were developed for the case of uniform stress with no body forces, it can be shown that identical results are found if body forces are included and the stresses are allowed to vary continuously. The total strain energy stored in an elastic solid occupying a region V is then given by the integral over the domain

$$U_T = \iiint U dxdydz \tag{5.24}$$

Using Hooke's law, the stresses can be eliminated from Eq. (5.20) and the strain energy can be expressed solely in terms of strain. For the isotropic case, this result becomes

$$U(\boldsymbol{\varepsilon}) = \frac{1}{2}\sigma_{ij}\varepsilon_{ij} = \frac{1}{2}\left(\lambda\varepsilon_{kk}\delta_{ij} + 2\mu\varepsilon_{ij}\right)\varepsilon_{ij} = \frac{1}{2}\lambda\varepsilon_{jj}\varepsilon_{kk} + \mu\varepsilon_{ij}\varepsilon_{ij} = \frac{1}{2}\lambda\left(\varepsilon_x + \varepsilon_y + \varepsilon_z\right)^2$$
$$+ \mu\left(\varepsilon_x^2 + \varepsilon_y^2 + \varepsilon_z^2 + \frac{1}{2}\gamma_{xy}^2 + \frac{1}{2}\gamma_{yz}^2 + \frac{1}{2}\gamma_{zx}^2\right)$$

Likewise, the strains can be eliminated and the strain energy can be written in terms of stress

$$U(\boldsymbol{\sigma}) = \frac{1+\nu}{2E}\sigma_{ij}\sigma_{ij} - \frac{\nu}{2E}\sigma_{jj}\sigma_{kk}$$
$$= \frac{1+\nu}{2E}\left(\sigma_x^2 + \sigma_y^2 + \sigma_z^2 + 2\tau_{xy}^2 + 2\tau_{yz}^2 + 2\tau_{zx}^2\right) - \frac{\nu}{2E}\left(\sigma_x + \sigma_y + \sigma_z\right)^2$$

After reviewing the various developed forms in terms of the stresses or strains, it is observed that the strain energy is a positive definite quadratic form with the property $U \geq 0$ for all values of σ_{ij} and ε_{ij}, with equality only for the case with $\sigma_{ij} = 0$ or $\varepsilon_{ij} = 0$. Actually, Eq. (5.23) is valid for all elastic materials, including both isotropic and anisotropic solids.

For the uniaxial deformation case, using Eq. (5.20) the derivative of the strain energy in terms of strain can be obtained as

$$\frac{\partial U(\boldsymbol{\varepsilon})}{\partial \varepsilon_x} = \frac{\partial}{\partial \varepsilon_x}\left(\frac{E\varepsilon_x^2}{2}\right) = E\varepsilon_x = \sigma_x \tag{5.25}$$

and likewise

$$\frac{\partial U(\boldsymbol{\sigma})}{\partial \sigma_x} = \frac{\partial}{\partial \sigma_x}\left(\frac{\sigma_x^2}{2E}\right) = \frac{\sigma_x}{E} = \varepsilon_x \tag{5.26}$$

These specific uniaxial results can be generalized for the 3D case, giving the relations

$$\sigma_{ij} = \frac{\partial U(\boldsymbol{\varepsilon})}{\partial \varepsilon_{ij}}, \varepsilon_{ij} = \frac{\partial U(\boldsymbol{\sigma})}{\partial \sigma_{ij}} \tag{5.27}$$

These results are again true for all elastic materials (isotropic and anisotropic). Thus, the strain energy function can be interpreted as playing a fundamental constitutive role in establishing general stress-strain relations for elastic materials. Such an approach in which the stresses are derivable from a strain energy function, that is, Eq. (5.27), is referred to as hyperelasticity. Note that this approach does not necessarily require that the relations between stress and strain to be linear, and thus this scheme is commonly used in the development of constitutive relations for nonlinear elastic solids. Only linear relations given by Hooke's law are incorporated in this textbook. Using Eq. (5.27), the following symmetry relations can be developed

$$\begin{cases} \dfrac{\partial \sigma_{ij}}{\partial \varepsilon_{kl}} = \dfrac{\partial \sigma_{kl}}{\partial \varepsilon_{ij}} \\ \dfrac{\partial \varepsilon_{ij}}{\partial \sigma_{kl}} = \dfrac{\partial \varepsilon_{kl}}{\partial \sigma_{ij}} \end{cases} \tag{5.28}$$

Going back to the general constitutive form $\sigma_{ij} = C_{ijkl}\varepsilon_{kl}$, Eq. (5.28) can be used to develop the additional symmetry relations

$$C_{ijkl} = C_{klij}$$

5.5.2 Decomposition of Strain Energy

The strain energy in an elastic solid may be decomposed into two parts, one associated with volumetric change U_v and the other caused by distortional (change in shape) deformation U_d.

$$U = U_v + U_d \tag{5.29}$$

The development of this decomposition is accomplished by using the definitions of spherical and deviatoric strain and stress tensors. For isotropic materials, the spherical stress produces only volumetric deformation, while the deviatoric stress causes only distortional changes. The volumetric strain energy is found by considering the spherical or hydrostatic components of stress and strain

$$U_v = \frac{1}{2}\tilde{\sigma}_{ij}\tilde{\varepsilon}_{ij} = \frac{1}{6}\sigma_{jj}\varepsilon_{kk} = \frac{1-2\nu}{6E}\sigma_{jj}\sigma_{kk} = \frac{1-2\nu}{6E}\left(\sigma_x + \sigma_y + \sigma_z\right)^2 \tag{5.30}$$

The distortional strain energy results from the deviatoric components or can be easily determined using Eq. (5.23), (5.29), and (5.30) to get

$$U_d = \frac{1}{12\mu}\left[\left(\sigma_x - \sigma_y\right)^2 + \left(\sigma_y - \sigma_z\right)^2 + \left(\sigma_z - \sigma_x\right)^2 + 6\left(\tau_{xy}^2 + \tau_{yz}^2 + \tau_{zx}^2\right)\right] \tag{5.31}$$

Particular failure theories of solids incorporate the strain energy of distortion by proposing that material failure or yielding will initiate when U_d reaches a critical value. It can be shown that the distortional strain energy is related to the octahedral shear stress previously discussed.

5.5.3 *Bounds on Elastic Constants*

Strain energy concepts allow us to generate particular bounds on elastic constants. For the isotropic case, the following three stress states previously investigated is considered.

5.5.3.1 Uniaxial Tension

In the loading scenario of uniform uniaxial deformation in the x direction (Fig. 5.19), the stress tensor can be written as:

$$\sigma_{ij} = \begin{bmatrix} \sigma & 0 & 0 \\ 0 & 0 & 0 \\ 0 & 0 & 0 \end{bmatrix}$$

In this case, the strain energy can be worked out as:

$$U = \frac{1+\nu}{2E}\sigma^2 - \frac{\nu}{2E}\sigma^2 = \frac{\sigma^2}{2E} \tag{5.32}$$

Because the strain energy is positive definite, the equation above implies that the modulus of elasticity must be positive, i.e.,

$$E > 0$$

5.5.3.2 Simple Shear

In the loading scenario of uniform simple shear (Fig. 5.20), the stress tensor can be written as:

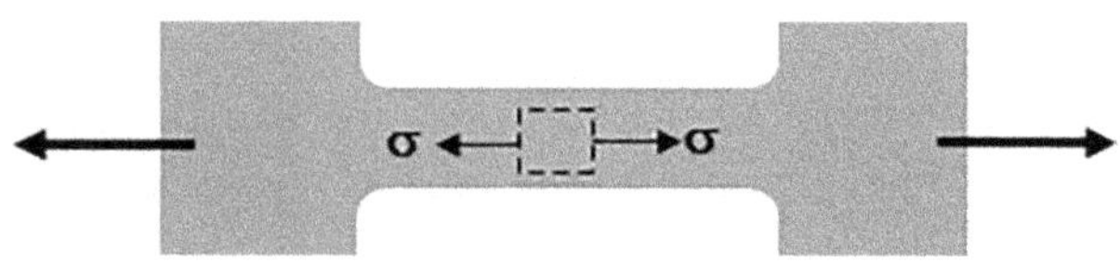

Fig. 5.19 Uniform uniaxial deformation

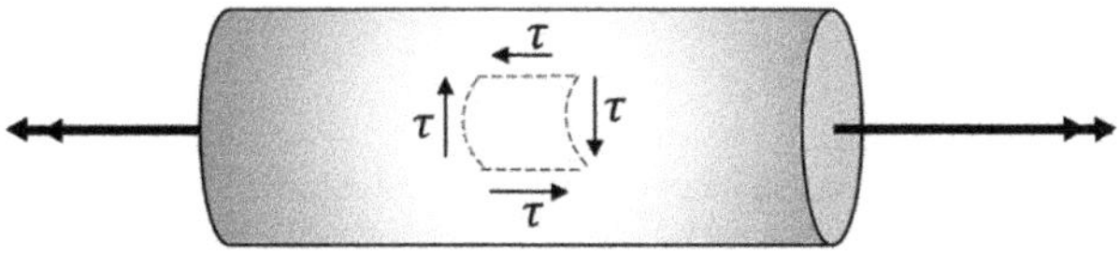

Fig. 5.20 Uniform simple shear loading

$$\sigma_{ij} = \begin{bmatrix} 0 & \tau & 0 \\ \tau & 0 & 0 \\ 0 & 0 & 0 \end{bmatrix}$$

The strain energy can be worked out as:

$$U = \frac{1+\nu}{2E}\left(2\tau^2\right) = \frac{\tau^2}{E}(1+\nu) \tag{5.33}$$

Again, because the strain energy is positive definite, the equation above gives:

$$1+\nu>0 \Rightarrow \nu>-1 \tag{5.34}$$

5.5.3.3 Hydrostatic Compression

The final loading scenario chosen is the uniform hydrostatic compression in which the stress tensor is given by

$$\sigma_{ij} = \begin{bmatrix} -p & 0 & 0 \\ 0 & -p & 0 \\ 0 & 0 & -p \end{bmatrix}$$

Putting it into the strain energy formulation yields the following result:

$$U = \frac{1+\nu}{2E}3p^2 - \frac{\nu}{2E}(-3p)^2 = \frac{3p^2}{2E}(1-2\nu) \tag{5.35}$$

The positive definite property of the strain energy gives another bound of the Poisson's ratio, i.e.,

$$1-2\nu>0 \Rightarrow \nu<\frac{1}{2} \tag{5.36}$$

Combing Eq. (5.34) with Eq. (5.36) gives the upper and lower bounds on Poisson's ratio:

$$-1 < \nu < \frac{1}{2} \tag{5.37}$$

Using relations between the elastic constants given in Table 4.5, the previous results also imply that

$$k > 0, \mu > 0$$

Experimental evidence indicates that most engineering materials have positive values of Poisson's ratio, and thus $0 < \nu < 1/2$. This further implies that $\lambda > 0$.

5.6 Principle of Virtual Work

Based on work and energy principles, several additional solution methods can be developed. These represent alternatives to the analytical methods based on differential equations. The principle of virtual work provides the foundation for many of these methods, and thus we begin our study by establishing this principle. The virtual displacement of a material point is a fictitious displacement such that the forces acting on the point remain unchanged. The work done by these forces during the virtual displacement is called the *virtual work*. For an object in static equilibrium, the virtual work is zero because the resultant force vanishes on every portion of an equilibrated body. The converse is also true that if the virtual work is zero, then the material point must be in equilibrium.

Let us introduce the following notational scheme. The virtual displacements of an elastic solid are denoted by $\delta u_i = \{\delta u, \delta v, \delta w\}$, and the corresponding virtual strains are then expressible as $\delta \varepsilon_{ij} = 1/2(\delta u_{i,\,j} + \delta u_{j,\,i})$. The standard elasticity boundary-value problem of a solid in equilibrium under the action of surface tractions over the boundary S_t with displacement conditions over the remaining boundary S_u is considered. Now imagine that the body undergoes a virtual displacement δu_i from its equilibrium configuration. The virtual displacement is arbitrary except that it must not violate the kinematic displacement boundary condition, and thus $\delta u_i = 0$ on S_u.

The virtual work done by the surface and body forces can be written as

$$\delta W = \int_{S_t} T_i^n \delta u_i dS + \int_V F_i \delta u_i dV \tag{5.38}$$

Now, because the virtual displacement vanishes on S_u, the integration domain of the first integral can be changed to S. Following standard procedures, this surface integral can be changed to a volume integral and combined with the body force term. These steps are summarized as

$$\begin{aligned} \delta W &= \int_S T_i^n \delta u_i dS + \int_V F_i \delta u_i dV \\ &= \int_S \sigma_{ij} n_j \delta u_i dS + \int_V F_i \delta u_i dV \\ &= \int_V \left(\sigma_{ij}\delta u_i\right)_{,j} dV + \int_V F_i \delta u_i dV \\ &= \int_V \left(\sigma_{ij,j}\delta u_i + \sigma_{ij}\delta u_{i,j}\right) dV + \int_V F_i \delta u_i dV \\ &= \int_V \left(-F_i \delta u_i + \sigma_{ij}\delta \varepsilon_{ij}\right) dV + \int_V F_i \delta u_i dV \\ &= \int_V \sigma_{ij}\delta\varepsilon_{ij} dV \end{aligned} \tag{5.39}$$

Now the last line in Eq. (5.39) is actually the virtual strain energy within the solid:

$$\begin{aligned} &\int_V \sigma_{ij}\delta\varepsilon_{ij} dV \\ &= \int_V \left(\sigma_x \delta\varepsilon_x + \sigma_y \delta\varepsilon_y + \sigma_z \delta\varepsilon_z + \tau_{xy}\delta\gamma_{xy} + \tau_{yz}\delta\gamma_{yz} + \tau_{zx}\delta\gamma_{zx}\right) dV \end{aligned} \tag{5.40}$$

Notice that the virtual strain energy does not contain the factor of 1/2 found in the general equation (Eq. (5.40)). This fact occurs because the stresses are constant during the virtual displacement.

Under the assumption of the existence of a strain energy function expressed in terms of the strains

$$\sigma_{ij} = \frac{\partial U(\boldsymbol{\varepsilon})}{\partial \varepsilon_{ij}} \tag{5.41}$$

and thus Eq. (5.40) can be written as

$$\int_V \sigma_{ij}\delta\varepsilon_{ij} dV = \int_V \frac{\partial U}{\partial \varepsilon_{ij}} \delta\varepsilon_{ij} dV = \delta \int_V U dV \tag{5.42}$$

Because the external forces are unchanged during the virtual displacements and the region V is fixed, the operator δ can be placed before the integrals in Eq. (5.38). Combining this with Eq. (5.42) allows Eq. (5.39) to be written as

$$\delta\left(\int_V U dV - \int_{S_t} T_i^n u_i dS - \int_V F_i u_i dV\right) = \delta(U_T - W) = 0 \tag{5.43}$$

This is one of the statements of the principle of virtual work for an elastic solid. The quantity $(U_T - W)$ actually represents the total potential energy of the elastic solid, and thus Eq. (5.43) states that the change in potential energy during a virtual displacement from equilibrium is zero. It should be noted that this principle is valid for all elastic materials including both linear and nonlinear stress-strain behaviors. The principle of virtual work provides a convenient method for deriving equilibrium equations and associated boundary conditions for various special theories of elastic bodies, including rods, beams, plates, and shells. In fact, even the continuum equations previously developed can be re-established using this method.

To illustrate the process of using the principle of virtual work to re-derive the basic equilibrium equations and related boundary conditions for the general elasticity problem, we can start with Eqs. (5.38) and (5.39) to have

$$\int_V \sigma_{ij}\delta\varepsilon_{ij}dV - \int_S T_i^n \delta u_i dS - \int_V F_i \delta u_i dV = 0 \tag{5.44}$$

The integrand of the first term can be reduced as

$$\sigma_{ij}\delta\varepsilon_{ij} = \frac{1}{2}\sigma_{ij}\left(\delta u_{i,j} + \delta u_{j,i}\right) = \sigma_{ij}\delta u_{i,j} = \left(\sigma_{ij}\delta u_i\right),j - \sigma_{ij,j}\delta u_i \tag{5.45}$$

and thus Eq. (5.44) can be expressed as

$$\begin{gathered}\int_V \left[\left(\sigma_{ij}\delta u_i\right)_{,j} - \sigma_{ij,j}\delta u_i\right]dV - \int_S T_i^n \delta u_i dS - \int_V F_i \delta u_i dV = 0 \\ \int_V \left(\sigma_{ij,j} + F_i\right)\delta u_i dV + \int_S \left(T_i^n - \sigma_{ij}n_j\right)\delta u_i dS = 0\end{gathered} \tag{5.46}$$

where we have used the divergence theorem to convert the volume integral $\int_V(\sigma_{ij}\delta u_i), jdV$ to a surface integral over S. For arbitrary δu_i, Eq. (5.46) is satisfied if

$$\sigma_{ij,j} + F_i = 0 \in V$$

and either

$$\delta u_i = 0 \in S_u$$

or

$$T_i^n = \sigma_{ij}n_j \in S_t$$

Conditions on S_u are commonly referred to as essential boundary conditions while those on S_t are called *natural boundary conditions*. Thus, we have demonstrated that the principle of virtual work can be used to generate equilibrium equations and associated boundary conditions for the general elasticity problem.

5.7 Principles of Minimum Potential and Complementary Energy

We now would like to use the results from the previous section to develop principles of minimum potential and complementary energy. Denoting the potential energy by $\Pi = U_T - W$, the virtual work statement indicated that the variation in potential energy from an equilibrium configuration was zero. Another way this is commonly stated is that potential energy is stationary in an equilibrium configuration. Such a condition implies that the potential energy will take on a local extremum (maximum or minimum) value for this configuration. It can be shown that the potential energy has a local minimum in the equilibrium configuration, and this leads to the following principle.

Principle of minimum potential energy: Of all displacements satisfying the given boundary conditions of an elastic solid, those that satisfy the equilibrium equations make the potential energy a local minimum.

An additional minimum principle can be developed by reversing the nature of the variation. Thus, the variation of the stresses is considered while holding the displacements constant. Let σ_{ij} be the actual stresses that satisfy the equilibrium equations and boundary conditions. Now consider a system of stress variations or virtual stresses $\delta\sigma_{ij}$ that also satisfies the stress boundary conditions (with $\delta T_i^n = \delta\sigma_{ij}n_j$ on S_t) and equilibrium equations with body force δF_i. In contrast to the previous development, we now investigate the complementary virtual work

$$\delta W^c = \int_S u_i \delta T_i^n dS + \int_V u_i \delta F_i dV \tag{5.47}$$

Employing the usual reduction steps as given in Eq. (5.39), the complementary virtual work statement can be reduced to

$$\int_S u_i \delta T_i^n dS + \int_V u_i \delta F_i dV = \int_V \varepsilon_{ij} \delta\sigma_{ij} dV \tag{5.48}$$

and the integral on the right-hand side is referred to as the complementary virtual strain energy. Introducing the complementary strain energy density function U^c, which is taken as a function of the stresses,

$$\varepsilon_{ij} = \frac{\partial U^c(\boldsymbol{\sigma})}{\partial \sigma_{ij}} \tag{5.49}$$

Using this result, the right-hand side of Eq. (5.48) can be expressed as

$$\int_V \varepsilon_{ij}\delta\sigma_{ij}dV = \int_V \frac{\partial U^c}{\partial \sigma_{ij}}\delta\sigma_{ij}dV = \delta\int_V U^c dV \tag{5.50}$$

Because the displacements do not vary and the region V is fixed, the operator δ can be placed before the integrals in Eq. (5.47). Combining this with Eq. (5.50) allows Eq. (5.48) to be written as

$$\delta\left(\int_V U^c dV - \int_{S_t} u_i T_i^n dS - \int_V u_i F_i dV\right) = \delta\left(U_T^c - W^c\right) = 0 \tag{5.51}$$

and thus, the variation in total complementary energy $\Pi^c = U^c - W^c$ is also zero in an equilibrium configuration. As before, it can be shown that this extremum in the complementary energy corresponds to a local minimum, thus leading to the following principle.

Principle of minimum complementary energy: Of all elastic stress states satisfying the given boundary conditions, those that satisfy the equilibrium equations make the complementary energy a local minimum.

Each of the previously developed minimum principles used general constitutive relations, either Eq. (5.41) or Eq. (5.49), and thus both principles are valid for all elastic materials regardless of whether the stress-strain law is linear or nonlinear. Fundamentally, the strain energy is expressed in terms of strain, while the complementary energy is functionally written in terms of stress. As shown in Fig. 5.21a, the strain energy for uniaxial deformation is equal to the area under the stress-strain curve, and thus $dU = \sigma\delta\varepsilon$. On the other hand, the complementary energy may be

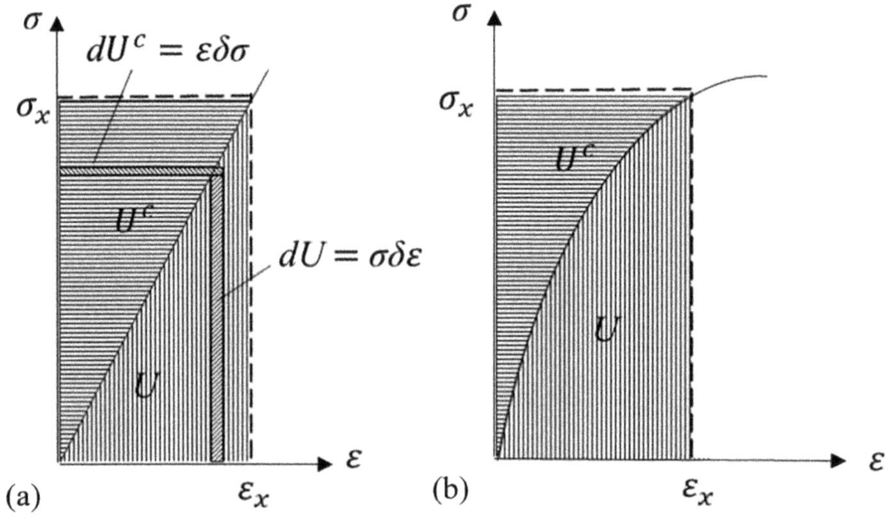

Fig. 5.21 Strain and complementary energy for linear and nonlinear elastic materials: (**a**) linear elastic case; (**b**) nonlinear elastic case

expressed by $dU^c = \varepsilon\delta\sigma$, and thus U^c will be the area *above* the uniaxial stress-strain curve as shown in Fig. 5.21a. For the uniaxial case with linear elastic behavior

$$U^c = \sigma_x\varepsilon_x - U = \frac{\sigma_x^2}{E} - \frac{\sigma_x^2}{2E} = \frac{\sigma_x^2}{2E} = \frac{1}{2}\sigma_x\varepsilon_x = U \tag{5.52}$$

This result is true for all deformations, and thus for linear elastic materials, the complementary energy is equal to the strain energy. Note also for this case

$$\frac{\partial U^c}{\partial \sigma_x} = \frac{\partial}{\partial \sigma_x}\left(\frac{\sigma_x^2}{2E}\right) = \frac{\sigma_x}{E} = \varepsilon_x \tag{5.53}$$

For the nonlinear elastic case, as shown in Fig. 5.21b, it is apparent that the strain energy and complementary energy will not be the same; that is, $U^c \neq U$. However, using the fact that $U^c = \sigma_x\varepsilon_x - U$, it follows that

Example 5.3: Euler-Bernoulli Beam Theory

Question: In order to demonstrate the utility of energy principles, an application dealing with the bending of an elastic beam, as shown in Fig. 5.22, is considered. The external distributed loading q will induce internal bending moments M and shear forces V at each section of the beam. According to classical Euler-Bernoulli theory, the bending stress σ_x and moment-curvature and moment-shear relations are given by

$$\sigma_x = -\frac{My}{I}, M = EI\frac{d^2w}{dx^2}, V = \frac{dM}{dx} \tag{5.54}$$

where $I = \iint_A y^2 dA$ is the area moment of inertia of the cross-section about the neutral axis, and w is the beam deflection (positive in y direction).

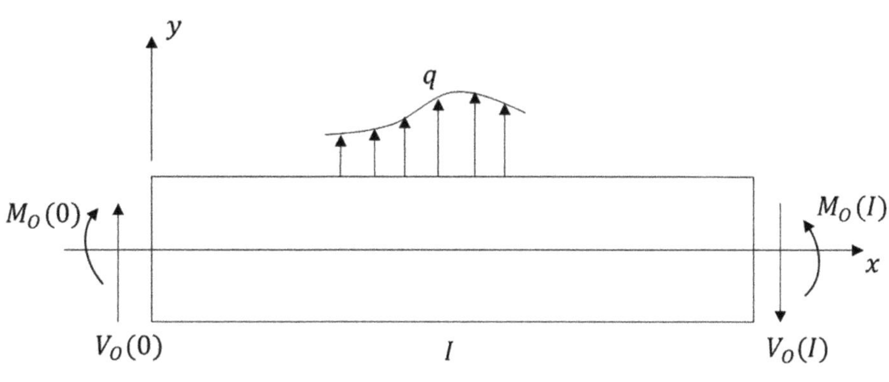

Fig. 5.22 Euler-Bernoulli beam geometry

$$\begin{aligned}\frac{\partial U^c}{\partial \sigma_x} &= \varepsilon_x + \sigma_x \frac{\partial \varepsilon_x}{\partial \sigma_x} - \frac{\partial U}{\partial \varepsilon_x}\frac{\partial \varepsilon_x}{\partial \sigma_x} \\ &= \varepsilon_x + \sigma_x \frac{\partial \varepsilon_x}{\partial \sigma_x} - \sigma_x \frac{\partial \varepsilon_x}{\partial \sigma_x} = \varepsilon_x\end{aligned} \tag{5.55}$$

Solution: Considering only the strain energy caused by the bending stresses

$$U = \frac{\sigma_x^2}{2E} = \frac{M^2 y^2}{2EI^2} = \frac{E}{2}\left(\frac{d^2 w}{dx^2}\right)^2 y^2 \tag{5.56}$$

and thus, the total strain energy in a beam of length l is

$$U_T = \int_0^l \left[\iint_A \frac{E}{2}\left(\frac{d^2 w}{dx^2}\right)^2 y^2 dA\right] dx = \int_0^l \frac{EI}{2}\left(\frac{d^2 w}{dx^2}\right)^2 dx \tag{5.57}$$

Now the work done by the external forces (tractions) includes contributions from the distributed loading q and the loadings at the ends $x = 0$ and l

$$W = \int_0^l qw dx - \left[V_0 w - M_0 \frac{dw}{dx}\right]_0^l \tag{5.58}$$

Therefore, the total potential energy for this beam case is given by

$$\Pi = U_T - W = \int_0^l \left[\frac{EI}{2}\left(\frac{d^2 w}{dx^2}\right)^2 - qw\right] dx + \left[V_0 w - M_0 \frac{dw}{dx}\right]_0^l \tag{5.59}$$

The first variation of this quantity must vanish

$$\begin{aligned}\delta\Pi &= \delta \int_0^l \left[\frac{EI}{2}\left(\frac{d^2 w}{dx^2}\right)^2 - qw\right] dx + \delta\left[V_0 w - M_0 \frac{dw}{dx}\right]_0^l \\ &= \int_0^l \delta\left[\frac{EI}{2}\left(\frac{d^2 w}{dx^2}\right)^2 - qw\right] dx + \left[V_0 \delta w - M_0 \frac{d\delta w}{dx}\right]_0^l \\ &= \frac{EI}{2}\int_0^l 2\frac{d^2 w}{dx^2}\frac{d^2 \delta w}{dx^2} dx - \int_0^l q\delta w dx + \left[V_0 \delta w - M_0 \frac{d\delta w}{dx}\right]_0^l = 0\end{aligned} \tag{5.60}$$

Now the first integral term can be integrated by parts twice to get

$$\int_0^l \left(EI\frac{d^4w}{dx^4} - q\right)\delta w dx + \left[\frac{d\delta w}{dx}(M - M_0) - \delta w(V - V_0)\right]_0^l = 0 \tag{5.61}$$

The integral and boundary terms must all vanish, thus implying

$$\begin{cases} \int_0^l \left(EI\frac{d^4w}{dx^4} - q\right)\delta w dx = 0 \\ V = V_0 \text{ or } \delta w = 0, x = 0, l \\ M = M_0 \text{ or } \delta\left(\frac{dw}{dx}\right) = 0, x = 0, l \end{cases} \tag{5.62}$$

For this integral to vanish for all variations δw, the fundamental lemma in the calculus of variations implies that the integrand must be zero, giving

$$EI\frac{d^4w}{dx^4} - q = 0 \tag{5.63}$$

This result is simply the differential equilibrium equation for the beam, and thus the stationary value for the potential energy leads directly to the governing equilibrium equation in terms of displacement and the associated boundary conditions. Of course, this entire formulation is based on the simplifying assumptions found in Euler-Bernoulli beam theory, and resulting solutions would not match with the more exact theory of elasticity results.

5.8 Rayleigh-Ritz Method

The previous beam example indicates a correspondence between the governing differential equation(s) and a variational problem of minimizing the potential energy of the system. Such a correspondence exists for many other types of problems in structural mechanics and elasticity. For problems of complicated shape or loading geometry, the solution to the governing differential equation cannot be found by analytical methods. For such cases, approximate solution schemes have been developed based on the variational form of the problem. Several such approximate schemes have been constructed, and one of the more important techniques is the Rayleigh-Ritz method.

This particular technique is based on the idea of constructing a series of trial approximating functions that satisfy the boundary conditions but not the differential equation(s). For the elasticity displacement formulation, this concept would express the displacements in the form

$$\begin{cases} u = u_0 + a_1u_1 + a_2u_2 + a_3u_3 + \ldots = u_0 + \sum_{j=1}^{N} a_ju_j \\ v = v_0 + b_1v_1 + b_2v_2 + b_3v_3 + \ldots = v_0 + \sum_{j=1}^{N} b_jv_j \\ w = w_0 + c_1w_1 + c_2w_2 + c_3w_3 + \ldots = w_0 + \sum_{j=1}^{N} c_jw_j \end{cases} \tag{5.64}$$

where the functions u_0, v_0, w_0 are chosen to satisfy any nonhomogeneous displacement boundary conditions and u_j, v_j, w_j satisfy the corresponding homogeneous boundary conditions. Note that these forms are not required to satisfy the traction boundary conditions. Normally, these trial functions are chosen from some combination of elementary functions such as polynomial, trigonometric, or hyperbolic forms. The unknown constant coefficients a_j, b_j, c_j are to be determined so as to minimize the potential energy of the problem, thus approximately satisfying the variational formulation of the problem under study. Using this type of approximation, the total potential energy will thus be a function of these unknown coefficients

$$\Pi = \Pi\left(a_j, b_j, c_j\right) \tag{5.65}$$

and the minimizing condition can be expressed as a series of expressions

$$\frac{\partial \Pi}{\partial a_j} = 0, \frac{\partial \Pi}{\partial b_j} = 0, \frac{\partial \Pi}{\partial c_j} = 0 \tag{5.66}$$

This set forms a system of 3N algebraic equations that can be solved to obtain the parameters a_j, b_j, c_j. Under suitable conditions on the choice of trial functions (completeness property), the approximation will improve as the number of included terms is increased.

Commonly, this technique is applied to a reduced elasticity problem involving only one or two components of displacement typically found in rods, beams, plates, and shells. Once the approximate displacement solution is obtained, the strains and stresses can be calculated from the appropriate field equations. However, since the strains and stresses are derivable through differentiation, the accuracy in these variables will in general not be as good as that obtained for the displacements. In order to demonstrate the Ritz technique, the Euler-Bernoulli beam problem is used again.

Example 5.4: Rayleigh-Ritz Solution of a Simply Supported Euler-Bernoulli Beam

Question: Considering a simply supported Euler-Bernoulli beam of length l carrying a uniform loading q_0, this 1D problem has displacement boundary conditions

$$w=0 \text{ at } x=0, l \tag{5.67}$$

and tractions or moment conditions

$$EI\frac{d^2w}{dx^2}=0 \text{ at } x=0, l \tag{5.68}$$

The Ritz approximation for this problem is of the form

$$w=w_0+\sum_{j=1}^{N} c_j w_j \tag{5.69}$$

Solution: With no nonhomogeneous boundary conditions, $w_0=0$. For this example, we choose a polynomial form for the trail solution. An appropriate choice that satisfies the homogeneous conditions (Eq. (5.67)) is $w_j = x^j(l-x)$. Note this form does not satisfy the traction conditions (Eq. (5.68)). Using the previously developed relation for the potential energy (Eq. (5.59)), we get

$$\begin{aligned}\Pi &= \int_0^l \left[\frac{EI}{2}\left(\frac{d^2w}{dx^2}\right)^2 - q_0 w\right] dx \\ &= \int_0^l \left[\frac{EI}{2}\left(\sum_{j=1}^{N} c_j\left[j(j-1)lx^{j-2} - j(j+1)x^{j-1}\right]\right)^2 - q_0\sum_{j=1}^{N} c_j x^j(l-x)\right] dx\end{aligned} \tag{5.70}$$

Retaining only a two-term approximation ($N=2$), the coefficients are found to be

$$c_1=\frac{q_0 l^2}{24EI}, c_2=0 \tag{5.71}$$

and this gives the following approximate solution:

$$w = \frac{q_0 l^2}{24EI} x(l - x) \tag{5.72}$$

Note that the approximate solution predicts a parabolic displacement distribution, while the exact solution to this problem is given by the cubic relation:

$$w = \frac{q_0 x}{24EI} \left(l^3 + x^3 - 2lx^2\right) \tag{5.73}$$

Actually, for this special case, the exact solution can be obtained from a Ritz scheme by including polynomials of degree three.

Other similar approximate techniques have been developed based on variational principles of complementary energy or the Reissner mixed formulation. A more generalized approximating scheme called the *weighted residual method* includes Ritz and several other techniques within the general approach. Although these approximate variational methods offer the potential to solve complex problems of engineering interest, they suffer a very important drawback involved with the selection of the approximating functions. Apart from the general properties the functions are required to satisfy, there exists no systematic procedure of constructing them. The selection process becomes more difficult when the domain is geometrically complex and/or the boundary conditions are complicated. Thus, these schemes have had limited success in solving such complicated problems. However, because these methods can easily provide approximate solutions over domains of simple shape with predetermined boundary conditions, they are ideally suited for finite element techniques, whereby a geometrically complex domain is divided into subdomains of simple geometry. Over each subdomain or element, the governing differential equation may be formulated using variational methods, and the approximation functions can then be systematically generated for each typical element.

Chapter 6
Two Dimensional Problems

Yongtao Lyu and Yonggang Zheng

6.1 Introduction

The equations presented in the previous chapters, e.g., the constitutive relation, are in the 3D space. To solve the engineering problems, these 3D formulations have to be solved, which is complex in some scenarios. It should be noted in some scenarios, the complex 3D problems can be simplified into 2D problems and consequently the calculations can be simpler and easier. In this chapter, the conditions when the 3D problems can be simplified into 2D problems are explained and the formulations related to plane elasticity are discussed.

6.2 Plane Stress Problem

6.2.1 The Simplification Conditions

The plane stress problem is defined as the scenario when the stress only occurs in a plane and the stress component perpendicular to this plane is zero. Under the following conditions, the 3D problem can be simplified as a plane stress problem.

(a) A very thin deformable body;
(b) The forces are parallel to the plane of the deformable body;
(c) The forces are evenly distributed along the thickness direction of the deformable body.

Y. Lyu (✉) · Y. Zheng
School of Mechanics and Aerospace Engineering, Dalian University of Technology, Dalian, Liaoning, China
e-mail: Yongtaolu@dlut.edu.cn; zhengyg@dlut.edu.cn

Y. Lyu, Y. Zheng, *Theory of Elasticity*,
https://doi.org/10.1007/978-981-95-6043-1_6

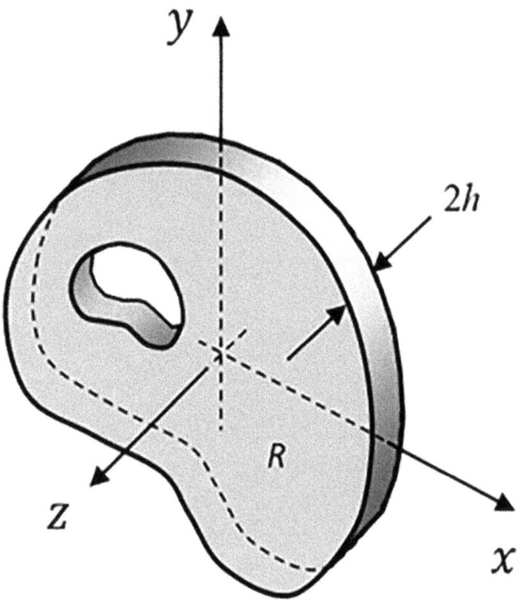

Fig. 6.1 Illustration of the plane stress problem

When the three conditions above are satisfied, the 3D problem can be simplified as a 2D problem and consequently the simplified formulations can be used to solve the original 3D problem. It should be noted that the first condition is given from the perspective of the dimension of the deformable body. The second and the third conditions are given from the perspective of the external force, i.e., the second condition refers to the direction of the force and the third one refers to the distribution pattern of the force. The three conditions are the necessary conditions for the simplification, which means if any one of them is not satisfied, the 3D problem cannot be simplified as a 2D problem.

Once the three conditions are satisfied, the thickness in the thin deformable body can be ignored and the problem can be solved using the following simplified equations. First, only three of the six stress components are not zeros, i.e., σ_x, σ_y, τ_{xy} in Fig. 6.1.

In summary, the stress tensor can be written as:

$$[\sigma] = \begin{bmatrix} \sigma_x & \tau_{xy} & 0 \\ \tau_{yx} & \sigma_y & 0 \\ 0 & 0 & 0 \end{bmatrix} \tag{6.1}$$

The non-zero stress components are independent of the z coordinate, i.e.,

$$\begin{cases} \sigma_x = \sigma_x(x, y, t) \\ \sigma_y = \sigma_y(x, y, t) \\ \tau_{xy} = \tau_{xy}(x, y, t) \end{cases} \tag{6.2}$$

6.2.2 The Strain Field

Now let's have a look at the constitutive equation for the plane stress problem. Considering the linear elastic constitutive equation:

$$\boldsymbol{\varepsilon} = -\frac{\nu}{E}\mathrm{Tr}(\boldsymbol{\sigma})\mathbf{1} + \frac{1+\nu}{E}\boldsymbol{\sigma} = -\frac{\nu}{E}\mathrm{Tr}(\boldsymbol{\sigma})\mathbf{1} + \frac{1}{2G}\boldsymbol{\sigma} \tag{6.3}$$

From which, the strain components can be worked out as:

$$\begin{cases} \varepsilon_x = \frac{1}{E}\left(\sigma_x - \nu(\sigma_y + \sigma_z)\right) = \frac{1}{E}\left(\sigma_x - \nu\sigma_y\right) & \gamma_{xy} = \frac{1}{G}\tau_{xy}, \\ \varepsilon_y = \frac{1}{E}\left(\sigma_y - \nu(\sigma_x + \sigma_z)\right) = \frac{1}{E}\left(\sigma_y - \nu\sigma_x\right) & \gamma_{xz} = \frac{1}{G}\tau_{xz} = 0, \\ \varepsilon_z = \frac{1}{E}\left(\sigma_z - \nu(\sigma_x + \sigma_y)\right) = -\frac{\nu}{E}\left(\sigma_x + \sigma_y\right) & \gamma_{yz} = \frac{1}{G}\tau_{yz} = 0, \end{cases} \tag{6.4}$$

where the conditions $\sigma_z = \tau_{xz} = \tau_{yz} = 0$ have been taken into account. From Eq. (6.4) it is concluded that the strains do not depend on the z-coordinate either, i.e., $\boldsymbol{\varepsilon} = \boldsymbol{\varepsilon}(x, y, t)$. In addition, the strain ε_z can be worked out from the other two normal strains as:

$$\varepsilon_z = -\frac{\nu}{1-\nu}\left(\varepsilon_x + \varepsilon_y\right) \tag{6.5}$$

In short, the strain tensor for the plane stress case can be written as:

$$\boldsymbol{\varepsilon}(x, y, t) = \begin{bmatrix} \varepsilon_x & \frac{1}{2}\gamma_{xy} & 0 \\ \frac{1}{2}\gamma_{xy} & \varepsilon_y & 0 \\ 0 & 0 & \varepsilon_z \end{bmatrix} \tag{6.6}$$

with

$$\varepsilon_z = -\frac{\nu}{1-\nu}\left(\varepsilon_x + \varepsilon_y\right)$$

From Eq. (6.4), the stress components can be expressed using the strain components as below:

$$\begin{cases} \sigma_x = \dfrac{E}{(1-\nu^2)}\left[\varepsilon_x + \nu\varepsilon_y\right] \\ \sigma_y = \dfrac{E}{(1-\nu^2)}\left[\varepsilon_y + \nu\varepsilon_x\right] \\ \tau_{xy} = \dfrac{E}{2(1+\nu)}\gamma_{xy} \end{cases} \tag{6.7}$$

From which the constitutive relation for the plane stress case can be written as:

$$\begin{Bmatrix} \sigma_x \\ \sigma_y \\ \tau_{xy} \end{Bmatrix} = \frac{E}{(1-\nu^2)}\begin{bmatrix} 1 & \nu & 0 \\ \nu & 1 & 0 \\ 0 & 0 & \dfrac{1-\nu}{2} \end{bmatrix}\begin{Bmatrix} \varepsilon_x \\ \varepsilon_y \\ \gamma_{xy} \end{Bmatrix} \tag{6.8}$$

The equilibrium equation for the plane stress case is reduced to:

$$\frac{\partial \sigma_x}{\partial x} + \frac{\partial \tau_{xy}}{\partial y} + F_x = 0 \tag{6.9a}$$

$$\frac{\partial \tau_{xy}}{\partial x} + \frac{\partial \sigma_y}{\partial y} + F_y = 0 \tag{6.9b}$$

where, F_x and F_y are functions of the coordinate (x and y) and $F_z = 0$. Expressing these equilibrium equations in terms of the displacements yield the Navier equations for the plane stress case:

$$\begin{cases} \mu\nabla^2 u + \dfrac{E}{2(1-v)}\dfrac{\partial}{\partial x}\left(\dfrac{\partial u}{\partial x} + \dfrac{\partial v}{\partial y}\right) + F_x = 0 \\ \mu\nabla^2 v + \dfrac{E}{2(1-v)}\dfrac{\partial}{\partial y}\left(\dfrac{\partial u}{\partial x} + \dfrac{\partial v}{\partial y}\right) + F_y = 0 \end{cases} \tag{6.10}$$

To develop the strain compatibility relations for the plane stress case, the original three relations, presented in Eq. (6.4), involving the out-of-plane components ε_z are commonly neglected. Afterwards, the only remaining compatibility relation for plane stress is:

$$\frac{\partial^2 \varepsilon_x}{\partial y^2} + \frac{\partial^2 \varepsilon_y}{\partial x^2} = 2\frac{\partial^2 \varepsilon_{xy}}{\partial x \partial y} \tag{6.11}$$

Expressing the reduced strain compatibility relation in terms of stress gives the corresponding Beltrami-Michell equation for the plane stress case:

$$\nabla^2\left(\sigma_x + \sigma_y\right) = -(1+v)\left(\frac{\partial F_x}{\partial x} + \frac{\partial F_y}{\partial y}\right) \tag{6.12}$$

6.2.3 *The Displacement Field*

To work out the displacement field, the geometric equation is recalled, i.e.,

$$\boldsymbol{\varepsilon}(\boldsymbol{x},t)=\nabla^S\boldsymbol{u}(\boldsymbol{x},t)=\frac{1}{2}(\boldsymbol{u}\otimes\nabla+\nabla\otimes\boldsymbol{u}) \tag{6.13}$$

which can be decomposed into the following two groups:

6.2.3.1 Those Which Do Not Affect the Displacement $\boldsymbol{u}_z$

$$\begin{cases} \varepsilon_x(x,y,t)=\dfrac{\partial u_x}{\partial x} \\ \varepsilon_y(x,y,t)=\dfrac{\partial u_y}{\partial y} \\ \gamma_{xy}(x,y,t)=2\varepsilon_{xy}=\dfrac{\partial u_x}{\partial y}+\dfrac{\partial u_y}{\partial x} \end{cases} \tag{6.14}$$

After the integration operation, displacements in the x and y directions can be worked out as:

$$\begin{cases} u_x=u_x(x,y,t) \\ u_y=u_y(x,y,t) \end{cases} \tag{6.15}$$

It can be seen, both u_x and u_y are independent of the z-coordinate.

Those in which the displacement $\boldsymbol{u}_z$ intervenes

$$\begin{cases} \varepsilon_z(x,y,t)=\dfrac{\partial u_z}{\partial z}=-\dfrac{\nu}{1-\nu}(\varepsilon_x+\varepsilon_y) \\ \gamma_{xz}(x,y,t)=2\varepsilon_{xz}=\dfrac{\partial u_x}{\partial z}+\dfrac{\partial u_z}{\partial x}=0 \\ \gamma_{yz}(x,y,t)=2\varepsilon_{yz}=\dfrac{\partial u_y}{\partial z}+\dfrac{\partial u_z}{\partial y}=0 \end{cases} \tag{6.16}$$

The displacement in the z direction can be worked out by the integration of the equations above, which can be written as a function of x, y, z and t, i.e., $u_z(x,y,z,t)$. It can be seen in the plane stress problem the unknown variables related to the plane of analysis are:

$$\begin{cases} \boldsymbol{u}(x,y,t) = \begin{Bmatrix} u_x \\ u_y \end{Bmatrix} \\ \boldsymbol{\varepsilon}(x,y,t) = \begin{Bmatrix} \varepsilon_x \\ \varepsilon_y \\ \gamma_{xy} \end{Bmatrix} \\ \boldsymbol{\sigma}(x,y,t) = \begin{Bmatrix} \sigma_x \\ \sigma_y \\ \tau_{xy} \end{Bmatrix} \end{cases} \tag{6.17}$$

in which the additional unknowns with respect to the general problem are either null, e.g., the following stress and strain components.

$$\sigma_z = \tau_{xz} = \tau_{yz} = \gamma_{xz} = \gamma_{yz} = 0$$

or can be calculated independently, e.g., the normal strain in the z direction,

$$\varepsilon_z = -\frac{\nu}{1-\nu}\left(\varepsilon_x + \varepsilon_y\right)$$

or do not intervene in the reduced problem, e.g., the displacement in the z direction, $u_z(x, y, z, t)$

6.3 Plane Strain Problem

6.3.1 The Simplification Condition

The plane strain problem is defined as the scenario when the strain only occurs in a plane and the strain component perpendicular to this plane is zero. Under the following conditions, the 3D problem can be simplified as a plane strain problem (Fig. 6.2):

(a) A very long deformable body;
(b) The forces are parallel to the cross-section of the deformable body;
(c) The forces are evenly distributed along the length direction of the deformable body.

Similar to the description in the plane stress problem, these three conditions are the necessary conditions for simplifying the 3D problem into the 2D problem. Furthermore, the first condition is given from the perspective of the dimension of the deformable body, the second is given from the perspective of the force direction and the third is given from the perspective of the force distribution. Once the three conditions are satisfied, a cross-section of the 3D deformable body, parallel to the external force, can be taken and used to represent the original 3D problem.

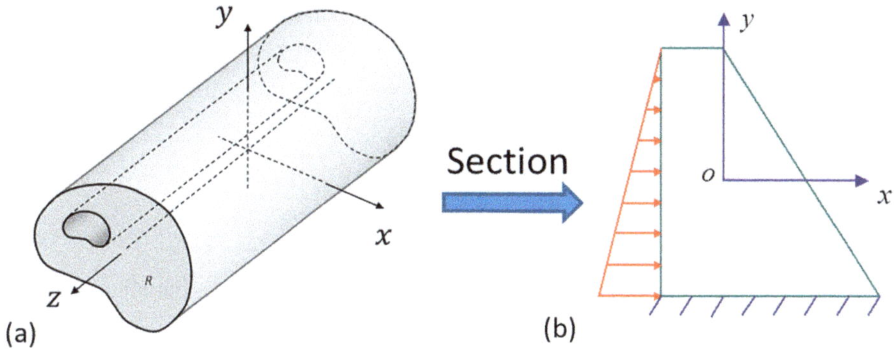

Fig. 6.2 Illustration of the plane strain problem: (**a**) The original 3D problem; (**b**) The simplified 2D plane strain problem

The displacement field for the plane strain case is:

$$\boldsymbol{u}=\left\{\begin{matrix} u_x \\ u_y \\ 0 \end{matrix}\right\}=\left\{\begin{matrix} u_x(x,y,t) \\ u_y(x,y,t) \\ 0 \end{matrix}\right\} \tag{6.18}$$

It should be noted that the displacements in the x and y directions are independent of the z-coordinate.

6.3.2 The Strain and Stress Fields

The strain field corresponding to the displacement field for a plane strain case is:

$$\begin{cases} \varepsilon_x(x,y,t)=\dfrac{\partial u_x(x,y,t)}{\partial x} \\ \varepsilon_y(x,y,t)=\dfrac{\partial u_y(x,y,t)}{\partial y} \\ \gamma_{xy}(x,y,t)=\dfrac{\partial u_x(x,y,t)}{\partial y}+\dfrac{\partial u_y(x,y,t)}{\partial x} \end{cases} \tag{6.19a}$$

$$\begin{cases} \varepsilon_z=\dfrac{\partial u_z}{\partial z}=0 \\ \gamma_{xz}(x,y,t)=\dfrac{\partial u_x(x,y,t)}{\partial z}+\dfrac{\partial u_z}{\partial x}=0 \\ \gamma_{yz}(x,y,t)=\dfrac{\partial u_y(x,y,t)}{\partial z}+\dfrac{\partial u_z}{\partial y}=0 \end{cases} \tag{6.19b}$$

It is seen the strain components associated with the z direction, i.e., ε_z, γ_{xz} and γ_{yz} are zeros. In short, the strain tensor for the plane strain case is:

$$\boldsymbol{\varepsilon}(x,y,t)=\begin{bmatrix} \varepsilon_x & \frac{1}{2}\gamma_{xy} & 0 \\ \frac{1}{2}\gamma_{xy} & \varepsilon_y & 0 \\ 0 & 0 & 0 \end{bmatrix} \tag{6.20}$$

It should be noted that the name plane strain arises from the fact that all strains are contained in the x-y plane.

Regarding the stress field, let's recall the linear elastic constitutive equation:

$$\boldsymbol{\sigma}=\lambda\mathrm{Tr}(\boldsymbol{\varepsilon})\mathbf{1}+2\mu\boldsymbol{\varepsilon}=\lambda\mathrm{Tr}(\boldsymbol{\varepsilon})\mathbf{1}+2G\boldsymbol{\varepsilon} \tag{6.21}$$

Combined with the strain field in Eqs. (6.19a) and (6.19b), the stress components for the plane strain case can be worked out:

$$\begin{cases} \sigma_x=(\lambda+2\mu)\ \varepsilon_x+\lambda\varepsilon_y=\dfrac{E(1-\nu)}{(1+\nu)(1-2\nu)}\left[\varepsilon_x+\dfrac{\nu}{1-\nu}\varepsilon_y\right] \\ \sigma_y=(\lambda+2\mu)\ \varepsilon_y+\lambda\varepsilon_x=\dfrac{E(1-\nu)}{(1+\nu)(1-2\nu)}\left[\varepsilon_y+\dfrac{\nu}{1-\nu}\varepsilon_x\right] \\ \sigma_z=\lambda(\varepsilon_x+\varepsilon_y) \end{cases} \tag{6.22a}$$

$$\begin{cases} \tau_{xy}=\mu\gamma_x \\ \tau_{xz}=\mu\gamma_{xz}=0 \\ \tau_{yz}=\mu\gamma_{yz}=0 \end{cases} \tag{6.22b}$$

It is concluded from Eqs. (6.22a) and (6.22b) that the stress components do not depend on the z-coordinate either, i.e., $\boldsymbol{\sigma}=\boldsymbol{\sigma}(x,y,t)$. In addition, the normal stress in the z direction σ_z can be expressed using the other two normal stresses, i.e.,

$$\sigma_z=\lambda(\varepsilon_x+\varepsilon_y)=\frac{\lambda}{2(\lambda+\mu)}(\sigma_x+\sigma_y)=\nu(\sigma_x+\sigma_y) \tag{6.23}$$

In summary, the stress tensor for the plane strain case can be written as:

$$\boldsymbol{\sigma}(x,y,t)=\begin{bmatrix} \sigma_x & \tau_{xy} & 0 \\ \tau_{xy} & \sigma_y & 0 \\ 0 & 0 & \sigma_z \end{bmatrix} \tag{6.24}$$

with $\sigma_z=\nu(\sigma_x+\sigma_y)$

The constitutive equation for the plane strain case can be written as:

$$\begin{Bmatrix} \sigma_x \\ \sigma_y \\ \tau_{xy} \end{Bmatrix} = \frac{E}{(1+\nu)(1-2\nu)} \begin{bmatrix} 1-\nu & \nu & 0 \\ \nu & 1-\nu & 0 \\ 0 & 0 & \frac{1-2\nu}{2} \end{bmatrix} \begin{Bmatrix} \varepsilon_x \\ \varepsilon_y \\ \gamma_{xy} \end{Bmatrix} \tag{6.25}$$

In addition, the equilibrium equations for the plane strain case are reduced to

$$\begin{aligned} \frac{\partial \sigma_x}{\partial x} + \frac{\partial \tau_{xy}}{\partial y} + F_x = 0 \\ \frac{\partial \tau_{yx}}{\partial x} + \frac{\partial \sigma_y}{\partial y} + F_y = 0 \end{aligned} \tag{6.26}$$

Notice that the reduced equilibrium relations are identical to the plane stress results. Using Eqs. (6.19a), (6.19b), (6.22a), and (6.22b), the equilibrium relations can be expressed in terms of displacements, yielding the Navier equations for the plane strain case:

$$\begin{cases} \mu \nabla^2 u + (\lambda + \mu) \dfrac{\partial}{\partial x}\left(\dfrac{\partial u}{\partial x} + \dfrac{\partial v}{\partial y}\right) + F_x = 0 \\ \mu \nabla^2 v + (\lambda + \mu) \dfrac{\partial}{\partial y}\left(\dfrac{\partial u}{\partial x} + \dfrac{\partial v}{\partial y}\right) + F_y = 0 \end{cases} \tag{6.27}$$

where ∇^2 is the 2D Laplacian operator $\nabla^2 = \left(\frac{\partial^2}{\partial x^2}\right) + \left(\frac{\partial^2}{\partial y^2}\right)$. Notice that these relations are similar but not identical to the plane stress results. The strain compatibility condition for the plane strain case is reduced to

$$\frac{\partial^2 \varepsilon_x}{\partial y^2} + \frac{\partial^2 \varepsilon_y}{\partial x^2} = 2 \frac{\partial^2 \varepsilon_{xy}}{\partial x \partial y} \tag{6.28}$$

Notice that the reduced compatibility condition for the plane strain is identical to that found in plane stress. Expressing this relation in terms of stress gives the corresponding Beltrami-Michell equation for the plane strain case:

$$\nabla^2 (\sigma_x + \sigma_y) = -\frac{1}{1-\nu}\left(\frac{\partial F_x}{\partial x} + \frac{\partial F_y}{\partial y}\right) \tag{6.29}$$

Notice that this result is similar but not identical to the corresponding plane stress relation. Similarly, to the plane stress problem, for the plane strain case, the unknown variables related to the $x - y$ plane are:

$$\begin{cases} \boldsymbol{u}(x,y,t) = \begin{Bmatrix} u_x \\ u_y \end{Bmatrix} \\ \boldsymbol{\varepsilon}(x,y,t) = \begin{Bmatrix} \varepsilon_x \\ \varepsilon_y \\ \gamma_{xy} \end{Bmatrix} \\ \boldsymbol{\sigma}(x,y,t) = \begin{Bmatrix} \sigma_x \\ \sigma_y \\ \tau_{xy} \end{Bmatrix} \end{cases} \tag{6.30}$$

The additional unknowns with respect to the general problem are either null, e.g.,

$$u_z = 0$$

$$\varepsilon_z = \gamma_{xz} = \gamma_{yz} = \tau_{xz} = \tau_{yz} = 0$$

or can be obtained from other unknowns, e.g.,

$$\sigma_z = \nu(\sigma_x + \sigma_y)$$

6.4 Conversion Between Plane Stress and Plane Strain Problems

In the formulation developments of plane strain and plane stress, it is apparent that although unfortunately the two theories do not have identical governing equations, many relations are very similar. Note that each theory has identical equilibrium equations, and boundary condition specifications. Furthermore, each theory has similar Navier equations and compatibility relations. Focus our attentions on these similar relations, it is observed that the basic difference between these equations is simply some coefficients involving the elastic material constants. This leads to the idea that perhaps a simple change in elastic moduli would bring one set of relations into an exact match with the corresponding result from the other plane theory. This in fact is the case, and it is easily shown that through transformation of the elastic modulus E and Poisson's ratio v (Table 6.1), all plane stress problems can be transformed into the corresponding plane model and vice versa. Therefore, solving

Table 6.1 Transformation between plane stress and strain problems

	E	v
Plane stress to plane strain	$\frac{E}{1-v^2}$	$\frac{v}{1-v}$
Plane strain to plane stress	$\frac{E(1+2v)}{(1+v)^2}$	$\frac{v}{1+v}$

one type of plane problem automatically gives the other solution through a simple transformation of elastic modulus and Poisson's ratio in the final answer.

It should be noted that for the Poisson's ratio $\nu = 0$, plane strain and plane stress solutions are identical.

Bibliography

1. Martin H. Sadd. Elasticity: Theory, Applications and Numerics. Academic Press, 2nd Edition, 2009.
2. Mikhail Filonenko-Borodich. Theory of Elasticity. University Press of the Pacific. 2003.
3. H. Leipholz. Theory of Elasticity. Springer Press, 2014.
4. Jerrold E. Marsden and Thomas J.R. Hughes. Mathematical Foundations of Elasticity. Dover Publications Inc. 2003.
5. Phillip L Gould. Introduction to Linear Elasticity. Springer Berlin Heidelberg, 2018.
6. Humphrey Hardy. Engineering Elasticity: Elasticity with less Stress and Strain. Springer International Publishing. 2022.

Y. Lyu, Y. Zheng, *Theory of Elasticity*, https://doi.org/10.1007/978-981-95-6043-1

GPSR Compliance
The European Union's (EU) General Product Safety Regulation (GPSR) is a set of rules that requires consumer products to be safe and our obligations to ensure this.

If you have any concerns about our products, you can contact us on

ProductSafety@springernature.com

In case Publisher is established outside the EU, the EU authorized representative is:

Springer Nature Customer Service Center GmbH
Europaplatz 3
69115 Heidelberg, Germany

www.ingramcontent.com/pod-product-compliance
Ingram Content Group UK Ltd.
Pitfield, Milton Keynes, MK11 3LW, UK
UKHW020633260926
475829UK00003BA/54

* 9 7 8 9 8 1 9 5 6 0 4 2 4 *